Piezoele Materi and Devices

Applications in Engineering and Medical Sciences

Piezoelectric Materials and Devices

Applications in Engineering and Medical Sciences

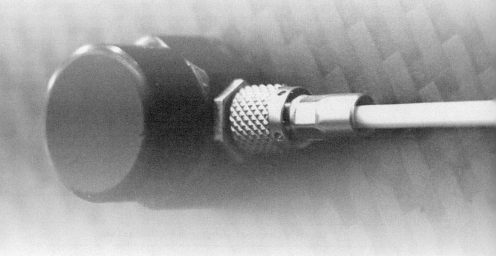

M. S. VIJAYA

CRC Press
Taylor & Francis Group
Boca Raton London New York

CRC Press is an imprint of the
Taylor & Francis Group, an **informa** business

CRC Press
Taylor & Francis Group
6000 Broken Sound Parkway NW, Suite 300
Boca Raton, FL 33487-2742

First issued in paperback 2017

ISBN-13: 978-1-4398-8786-8 (hbk)
ISBN-13: 978-1-138-07742-3 (pbk)

Library of Congress Cataloging-in-Publication Data

Vijaya, M. S.
    Piezoelectric materials and devices : applications in engineering and medical sciences / M.S. Vijaya.
        p. cm.
    Includes bibliographical references and index.
    ISBN 978-1-4398-8786-8 (hardback)
    1. Piezoelectric devices. I. Title.

TK7872.P54V55 2012
621.3815--dc23   2012016094

Dedicated to

My mother who is no more in this world but lives in my heart.

Contents

Preface

Although piezoelectric materials have been known for more than a century, the materials have gained importance over the last few decades because of their high potential as smart materials. The materials have inherent transducer characteristics, being capable of both sensing and actuation. They convert mechanical energy into electrical energy, which makes them useful as sensitive sensors of mechanical inputs such as pressure, strain, vibration, rotation, sound, ultrasound, etc. They also convert electrical energy to mechanical energy, which makes them useful as actuators. The materials have a large number of engineering and medical applications by virtue of these characteristics. They belong to the class of smart materials and are essential components of majority of smart systems such as industrial and medical robots.

The phenomenon of piezoelectricity was discovered as early as 1880 by Jaques and Pierri Curie in single crystals of quartz. They later reported piezoelectric behavior in other naturally occurring materials such as Rochelle salt, tourmaline, and topaz. The discovery of ferroelectric ceramics barium titanate, and lead zirconate titante (PZT) during the 1940s and 1950s led to a spate of research activities on these materials. Over the years, a wide variety of transducers, sensors, and actuators have been developed using the ceramic PZT, which is one of the most sensitive piezoelectric materials. Another important piezoelectric material that has created a lot of interest is the polymer polyvinylidene fluoride (PVDF), which was discovered in 1969. A polymer exhibiting transducer characteristics has special advantages over the ceramic because of the more flexible and less brittle nature of polymers. Zinc oxide is a relatively newfound piezoelectric material that has been used in nano-crystalline form for piezoelectric applications such as micro actuators and energy harvesting.

The field of piezoelectricity is of interest to professionals from diverse fields such as materials scientists, industrial design engineers, biomedical engineers, doctors, etc. Materials scientists have been trying to find piezoelectric materials of improved performances, for example, doped PZT, copolymers of PVDF, ceramic-polymer composite piezoelectrics, nano-crystals of zinc oxide. Design engineers are devising new innovative smart devices using these materials, and there have been a large number of patent applications of devices using piezoelectric materials. Biomedical engineers have developed many piezoelectric devices for medical and bioengineering applications. Some of the piezoelectric devices are routinely used by doctors in hospitals for medical diagnosis and therapy.

It can be said that there are no engineering fields in which piezoelectric materials have not found applications. Piezoelectric devices are being used in fields such as aerospace, automotive, industrial equipment, defense equipment, robotics, SONAR, household appliances, etc. Medical applications of piezoelectric devices are in drug delivery, surgery, diagnosis, and therapy.

The present book aims at introducing the reader to this fascinating class of materials. The treatment includes physics of piezoelectric materials and their characteristics and applications. The theory of piezoelectricity is explained in simple terms without expecting too much of background.

The first chapter introduces various types of dielectrics and their classification based on their characteristics. The characteristics of electrostrictive, piezoelectric, pyroelectric, and ferroelectric materials and their applications are discussed briefly.

The second chapter deals with the mathematical formulation of piezoelectric effects and the definition of various piezoelectric constants. Inherent characteristics of piezoelectric materials such as motors and generators are described with relevant equations. The structure and properties of practical piezoelectric materials such as quartz, lead zirconate titanate, barium titanate, zinc oxide, and polyvinylidene fluoride are described in detail in Chapter 3. Composite piezoelectric materials that have a special advantage over the single-phase piezoelectrics are also discussed.

Chapters 4 and 5 cover the entire gamut of piezoelectric devices used in engineering and medical applications. Commercial piezoelectric products such as PZT ultrasonic transducers and detectors, multilayer actuators, microsensors, and microactuators are discussed in detail. Some of the recent applications of piezoelectric materials such as ZnO nano fibers for energy harvesting and thin films of polyvinylidene fluoride for structural health monitoring are discussed briefly.

There has been a spate of new developments over the last 2 decades on biomedical applications of piezoelectric devices such as drug delivery, blood flow and blood pressure monitoring, robotic operating tools, etc. These devices are covered in Chapter 5.

In the last chapter (Chapter 6), design and virtual prototyping of piezoelectric devices using FE software tools are described. Case studies of some specific piezoelectric devices using the FE software tools ANSYS and PAFEC are included.

It is expected that the book will provide sufficient information to design engineers, scientists, and technologists to adopt piezoelectric materials in the development of smart devices for specific applications and motivate students of engineering and science to take up research in the area for developing innovative devices. The book could be used as a textbook or reference book by teachers and students who want to learn the fundamentals of piezoelectric materials and their applications.

I am thankful to all authors of books cited in the bibliography without which writing a book of this type would not have been possible. Of course, a wealth of information available on the internet has been of immense use as the field is highly specialized and books available to me were not sufficient to get all the information. Most of the references given under the applications chapters (Chapters 4 and 5) are the relevant web-sites. I wish to place on record my sincere thanks to Mr. K. Jagadeesh, Assistant Professor, M. S. Ramaiah School of Advanced Studies, who helped in finite element analyses of piezoelectric devices using finite element software tools included in Chapter 6.

I am grateful to the reviewers who gave positive comments on the initial proposal for the book and also gave useful suggestions for improvements. I have attempted to include most of their suggestions.

M. S. Vijaya

1

Introduction

1.1 Classification of Dielectric Materials

All dielectric materials when subjected to an external electric field undergo change in dimensions. This is due to the displacements of positive and negative charges within the material. A dielectric crystal lattice may be considered to be made up of cations and anions connected by springs (interionic chemical bonds). When an external electric field is applied to the material, the cations get displaced in the direction of the electric field and the anions get displaced in the opposite direction, resulting in net deformation of the material. The change in dimension may be very small or may be quite significant, depending on the crystal class to which the dielectric belongs.

Among the total of 32 crystal classes, 11 are centrosymmetric (i.e., possess a centre of symmetry or inversion centre) and 21 are noncentrosymmetric (do not possess a centre of symmetry).

When a dielectric material possessing a centre of symmetry is subjected to an external electric field, due to the symmetry (inversion centre), the movements of cations and anions are such that the extension and contraction get cancelled between neighbouring springs (chemical bonds) and the net deformation in the crystal is ideally nil. But the chemical bonds are not perfectly harmonic and, due to the anharmonicity of the bonds, there will be second-order effects resulting in a small net deformation of the lattice. The deformation in this case is proportional to the square of the electric field. That is, the deformation is independent of the direction of the applied electric field. The effect is called *electrostrictive effect*. The anharmonic effect exists in all dielectrics, and so it can be said that all dielectrics are electrostrictive.

When a dielectric material belonging to a noncentrosymmetric class (except the octahedral class) is subjected to an external electric field, there will be asymmetric movement of the neighbouring ions, resulting in significant deformation of the crystal and the deformation is directly proportional to the applied electric field. These materials exhibit an electrostrictive effect due to the anharmonicity of the bonds, but it is masked by the more significant

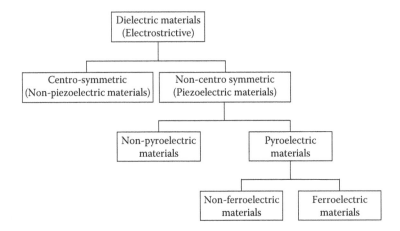

FIGURE 1.1
Classification of dielectric materials.

asymmetric displacement. The materials are called *piezoelectric materials*. The classification of dielectric materials based on their response to external stimuli is shown in Figure 1.1.

Each of these groups of materials exhibits certain special characteristics that make them important engineering materials. The materials belong to the class of ***smart materials*** because they exhibit inherent transducer characteristics. The characteristics of the materials are discussed in detail in the rest of the sections of the book.

1.2 Important Dielectric Parameters

1.2.1 Electric Dipole Moment p

In an atom or a molecule when the centres of positive and negative charges are separated by a certain distance d, the atom or the molecule possesses an **electric dipole moment** given by

$$\vec{p} = q\vec{d} \tag{1.1}$$

where q is the charge, and d is the separation between the positive and negative charge centres. **p** is a vector; the direction of the vector is from the negative charge to the positive charge, and its unit is Cm (coulomb meter).

1.2.2 Polar and Nonpolar Dielectric Materials

Dielectric materials may be classified as polar and nonpolar.

In nonpolar dielectric materials, normally the atoms do not possess an electric dipole moment as the centres of positive and negative charges coincide. Typical examples of nonpolar dielectrics are oxygen, nitrogen, benzene, methane, etc. When these materials are subjected to an external electric field, the centres of positive and negative charges get separated and thus dipole moments are induced. The induced dipole moments disappear once the electric field is removed.

In polar dielectric materials, each atom or molecule possesses a dipole moment as the centres of positive and negative charges do not coincide. Typical examples of polar dielectrics are water, HCl, alcohol, NH_3, etc. Most of the ceramics and polymers fall under this category. When an external electric field is applied to these materials, the electric dipoles tend to orient themselves in the direction of the field.

1.2.3 Electric Polarization P

A polar dielectric material consists of a large number of atoms or molecules each possessing an electric dipole moment. The total or the net dipole moment of the dielectric material is the vector sum of all the individual dipole moments given by

$$\sum_i \vec{p_i}$$

Electric polarization P is defined as the total dipole moment per unit volume and is given by

$$\vec{P} = \frac{\sum_i \vec{p_i}}{V} \tag{1.2}$$

where V is the volume of the material. The unit of P is C/m^2 (coulomb/meter^2), and it is sometimes called the *surface charge density*. P is a vector normal to the surface of the material.

Normally, in a polar dielectric, the individual electric dipoles are all randomly oriented and so the net polarization is zero. When an electric field is applied, the individual dipoles tend to orient themselves in the direction of the electric field and the material develops a finite polarization. The polarization increases with an increasing electric field and attains saturation when all the dipole moments are oriented in the direction of the field.

1.2.4 Dielectric Displacement or Flux Density *D*, Dielectric Constant ε_r, and Electric Susceptibilty χ

When an electric field *E* is applied to a dielectric material, the material develops a finite polarization *P* (induced polarization in nonpolar materials and orientation polarization in polar materials). The electric flux density *D* developed inside the material due to the external field *E* is given by

$$\vec{D} = \varepsilon_o \vec{E} + \vec{P} \tag{1.3}$$

where ε_o is the permittivity of free space.
    D is also expressed by the relation

$$\vec{D} = \varepsilon \vec{E} = \varepsilon_o \varepsilon_r \vec{E} \tag{1.4}$$

where ε_r is the *relative permittivity* or *dielectric constant* of the dielectric material. The unit of *D* is C/m^2 (same as that of *P*).
    The dielectric constant ε_r is also defined by the ratio

$$\varepsilon_r = \frac{D}{\varepsilon_o E}$$

ε_r is a unitless quantity and is always greater than 1.
    The polarization *P* is directly related to the applied electric field by the relation

$$\vec{P} = \varepsilon_o \chi \vec{E} \tag{1.5}$$

where ε_o is the permittivity of free space and χ is called the *electric susceptibility* of the material.
    From Equations 1.3, 1.4, and 1.5, we get the relation between the dielectric constant ε_r and electric susceptibility χ as

$$\varepsilon_r = 1 + \chi \tag{1.6}$$

1.2.5 Spontaneous Polarization P$_S$

The spontaneous polarization is the polarization that exists inherently in the material without the need for an external electric field. Pyroelectric and ferroelectric materials exhibit spontaneous polarization. In pyroelectric materials, the spontaneous polarization is not observed normally as the surface charges get compensated by the free charges in the surroundings. But the spontaneous polarization decreases when the material is heated, and so the change in polarization on heating can be observed (see Section 1.5). In ferroelectric materials, the direction of spontaneous polarization can be changed by applying an external electric field. Thus, in ferroelectric materials, the spontaneous polarization can be increased in a given direction by applying an electric field (see Section 1.6).

1.3 Electrostrictive Effect

All dielectrics exhibit an electrostrictive effect. Application of an external electric field polarizes the material. The process of polarization involves the orientation of the dipoles that results in deformation of the material. The strain x is related to the polarization P by

$$x = QP^2 \qquad (1.7)$$

where Q is the *electostrictive coefficient*. Using Equation 1.3 and noting that for materials with high dielectric constant $P \gg \varepsilon_o E$, we can write

$$x = QD^2 = Q\varepsilon^2 E^2 \qquad (1.8)$$

The previous equation describes the electrostrictive effect in which the resulting strain is proportional to the square of the electric field. Thus, the strain generated is independent of the polarity of the applied field. Since x is a second-rank tensor and E is a vector, the electrostrictive coefficient is a fourth-rank tensor. It relates the strain tensor x to the various cross products of the components of the electric field vector E or D. The tensor form of Equation 1.8 is

$$x_{ij} = Q_{ijkl} D_{km} D_{ml} \qquad (1.9)$$

Although all dielectrics exhibit an electrostrictive effect, it is more pronounced in some dielectrics that have special phase transition characteristics. The best-known electrostrictive materials for actuator applications are lead magnesium niobate (PMN) and lead magnesium niobate–lead titanate (PMN-PT).

PMN is cubic at high temperature. On annealing, it transforms to a partially ordered state. On further cooling, it passes through a diffuse phase transformation, where it is poised on instability with microregions fluctuating in polarization. In this phase, it exhibits large dielectric constant and electrostrictive coefficients. Much below room temperature, it transforms to a ferroelectric rhombohedral phase. It is useful as an electrostrictive actuator at room temperature in the nonferroelectric phase (paraelectric cubic phase). Typically, electrostrictive PMN ceramic shows a strain of 10^{-4} for an electric field of about 1 MV/m.

The advantages of electrostrictive ceramic over piezoelectric ceramic are that the impedance of the materials is much higher than the piezoelectric materials and they exhibit very little hysteresis.

1.4 Piezoelectric Effect

Dielectric materials that belong to the class of noncentrosymmetric crystals are classified as piezoelectric materials. When these materials are subjected to an external electric field, there will be asymmetric displacements of anions and cations that cause considerable net deformation of the crystal. The resulting strain is directly proportional to the applied electric field unlike electrostrictive materials in which the strain is proportional to E^2. The strain in a piezoelectric material is extensive or compressive, depending on the polarity of the applied field. This effect is called the *piezoelectric effect*, or to be more precise, *indirect piezoelectric effect*.

Piezoelectric materials exhibit another unique property; when they are subjected to external strain by applying pressure/stress, the electric dipoles in the crystal get oriented such that the crystal develops positive and negative charges on opposite faces, resulting in an electric field across the crystal. This is exactly the reverse of the above mentioned indirect piezoelectric effect. Jaques and Pierri Curie first observed this effect in quartz crystal in 1880 and called this *piezoelectricity;* "piezo" meaning pressure. The effect is called the *direct piezoelectric effect*.

The direct and indirect piezoelectric effects are illustrated in Figures 1.2 and 1.3. In the direct effect when a poled piezoelectric material (see Section 1.7) is subjected to tensile stress, in the direction parallel to the poling direction, a positive voltage is generated across the faces (Figure 1.2b). When the material is subjected to compressive stress in the direction, a negative voltage is generated across the faces (Figure 1.2c). In the indirect effect, when an external voltage is applied to the material, the material gets extended if the polarity of the voltage is the same as that of the field applied during poling (Figure 1.3b) and, when the voltage is applied in the reverse direction, the material gets compressed (Figure 1.3c).

Figure 1.4 shows the effect of an alternating field on a poled piezoelectric material. The alternating field makes the material to extend and contract alternately at the same frequency as the applied field. The vibration produces an acoustic field (sound or ultrasonic field) in the vicinity of the material. This effect is used for the generation of acoustic fields.

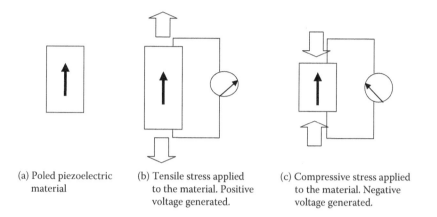

(a) Poled piezoelectric material

(b) Tensile stress applied to the material. Positive voltage generated.

(c) Compressive stress applied to the material. Negative voltage generated.

FIGURE 1.2
Direct piezoelectric effect: (a) Poled piezoelectric material. (b) When tensile stress is applied to the material, the material develops voltage across its face with the same polarity as the poling voltage. (c) When a compressive stress is applied to the material, the material develops voltage with polarity opposite to that of the poling voltage.

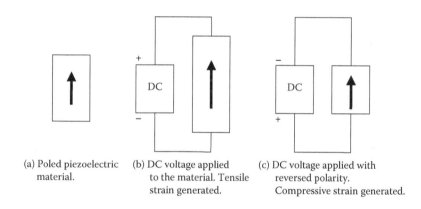

(a) Poled piezoelectric material.

(b) DC voltage applied to the material. Tensile strain generated.

(c) DC voltage applied with reversed polarity. Compressive strain generated.

FIGURE 1.3
Indirect piezoelectric effect: (a) Poled piezoelectric material. (b) When a DC field is applied with the same polarity as the poling field, the material develops tensile strain. (c) When a DC field is applied in the reverse direction, the material develops compressive strain.

The direct and the indirect piezoelectric effects have many applications as the effects involve conversion of mechanical energy into electrical energy and vice versa. The applications include generation and detection of ultrasonic waves, pressure sensors, and actuators. Ultrasonic is extensively used both in engineering and medical fields. In engineering, it is used in non-destructive testing of materials (NDT), underwater acoustics (SONAR), ultrasonic drilling, energy harvesting, etc., and in medical fields, it is used for diagnosis (sonography), therapy (drug delivery), and surgery. As sensors and

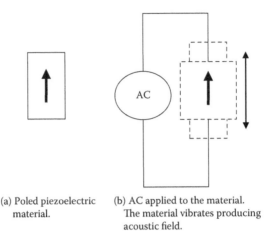

(a) Poled piezoelectric   (b) AC applied to the material.
material.   The material vibrates producing
                                 acoustic field.

FIGURE 1.4
Effect of AC field on a piezoelectric material: (a) Poled piezoelectric material; (b) AC field is applied to the material. The material gets extended and contracted alternately; that is, the material vibrates producing an acoustic field in the vicinity.

actuators, they have a wide variety of applications in both engineering and medical fields. Direct and indirect piezoelectric effects and their applications are discussed in great depth in the following chapters of the book.

1.5 Pyroelectric Effect

Pyroelectric materials exhibit spontaneous polarization, but their polarization cannot be reoriented by application of external electric fields. Normally, in these materials, the spontaneous polarization cannot be observed as the surface charges get compensated by free charges in the surroundings. But when the material is heated, the spontaneous polarization decreases and this change in polarization can be measured. They are used as sensors to detect infrared radiation.

The pyroelectric effect is described by the equation

$$dP_s = -pdT \tag{1.10}$$

where dT is the change in temperature, and dP_S is the corresponding change in the spontaneous polarization. p is the pyroelectric coefficient. The negative sign indicates that the spontaneous polarization decreases with increasing temperature. Since temperature is a scalar and polarization is a vector, the pyroelectric coefficient is a vector.

In practical applications, the pyroelectric material is subjected to a steady increase in temperature. The current generated is directly proportional to the temperature gradient. The current density is given by

$$J = -\frac{1}{A}\frac{dQ}{dt} = -\frac{dP_s}{dt} = -\frac{dP_s}{dT}\frac{dT}{dt} \tag{1.11}$$

Using Equation 1.10,

$$J = p\frac{dT}{dt} \tag{1.12}$$

The current generated is measured using an external circuit. Ferroelectric materials exhibit better pyroelectric effects than nonferroelectric materials. So, practical pyroelectric devices mostly use ferroelectric materials. The ferroelectric material is poled before using as a pyroelectric device such as an infrared detector. A thin poled ferroelectric wafer is used for infrared detection. Infrared radiation is modulated at a particular frequency using a chopper and made incident on the poled pyroelectric wafer. The alternating change in temperature generates an alternating current that is converted to DC and measured. This is illustrated in Figure 1.5.

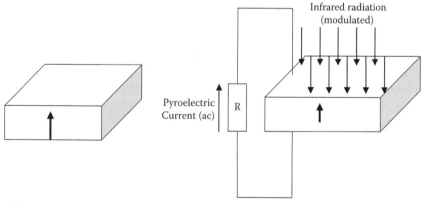

(a) Poled pyroelectric material

(b) Infrared radiation made incident on the material. The polarization of the material decreases, generating current through the close circuit.

FIGURE 1.5
When infrared radiation is incident on a poled pyroelectric material, the change in temperature generates an electric current through the material that is measured using an external circuit.

1.6 Ferroelectric Materials

Ferroelectric materials are a subclass of pyroelectric materials, which are again a subclass of piezoelectric materials. Thus, ferroelectric materials exhibit both piezoelectric and pyroelectric properties. Pyroelectric and piezoelectric materials that are ferroelectric have more sensitive characteristics, and so most of the practical pyroelectric and piezoelectric devices use ferroelectric materials.

Ferroelectric materials exhibit spontaneous polarization such as pyroelectrics, but they have the characteristic property of *reversible polarization*. That is, the spontaneous polarization can be reversed by applying an external electric field. Ferroelectrics are characterized by the following properties:

- Ferroelectric hysteresis
- Spontaneous polarization
- Reversible polarization
- Ferroelectric transition temperature

In a ferroelectric material, the polarization P as a function of the applied electric field E is nonlinear.

1.6.1 Ferroelectric Domains

Ferroelectric domain is defined as a small microscopic region in the material within which all the electric dipoles are oriented in the same direction due to a strong short-range interaction caused by internal electric fields. A ferroelectric material consists of a large number of domains with each domain having a specific polarization direction. Normally, the domains are all randomly oriented, and so the net polarization of the material is zero in the absence of an external electric field. When an external electric field is applied, the domains tend to get oriented in the direction of the applied field. That is, the domains that are in the direction of the external field grow in size at the expense of the other domains. As the external field is increased, more and more domains get oriented in the direction and ultimately the material ideally consists of a single domain.

1.6.2 Ferroelectric Hysteresis

The variation of polarization with an applied electric field in a ferroelectric material is shown in Figure 1.6.

Initially, when the applied field is zero, the ferroelectric domains are all randomly oriented and so the polarization is zero (point O). As the field is increased, the domains get oriented in the direction of the field, and the polarization increases linearly in the beginning. This is shown by the portion of the curve OA. As the field is further increased, more and more domains get

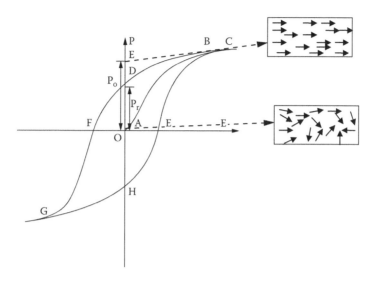

FIGURE 1.6
P versus E curve for a ferroelectric material.

oriented, the curve becomes nonlinear, and ultimately when all the domains get oriented, the polarization attains the maximum value (point B). The polarization at this point P_o (OE) is called *saturation polarization*. If the electric field is now reduced gradually, the polarization decreases but the curve is not retraced. The decrease in polarization is rather slow; that is, the polarization lags behind the electric field. This is shown by the portion BD. When the field is reduced to zero, there remains a finite polarization called the *remnant polarization* P_R (OD). In order to make the remnant polarization disappear, an electric field in the reverse direction has to be applied. At an electric field of $-E_C$ called the *coercive field*, the polarization becomes zero (at point F). If the field is further increased in the reverse direction beyond E_C, the domains get oriented in the direction of the field and the polarization increases with increasing field (in the new reversed direction). The polarization attains the maximum value, saturation polarization $(-P_S)$ in the reverse direction (at G). If the field is now again reduced back to zero, the curve traces the path GH and there will be a remnant polarization $-P_R$ (at H). If the field is increased from zero in the positive direction, the remnant polarization disappears when the field is $+E_C$. Further increase in the field will trace the path E_cB, closing the loop. The closed loop is called the *hysteresis curve*.

1.6.3 Ferroelectric Transition Temperature

Ferroelectric material loses its spontaneous polarization beyond a certain temperature. This temperature is called the *ferroelectric transition temperature*. At this temperature, the material is said to undergo a transition from ferroelectric phase to a paraelectric phase.

1.7 Poling

Ferroelectric materials can be poled. "Poling" is the process of generating net remnant polarization in the material by applying sufficiently high electric field.

When an electric field is applied to a ferroelectric material, the microscopic ferroelectric domains orient themselves in the direction of the applied field. As the electric field is increased, more and more domains get oriented and, at a sufficiently high electric field, almost all the domains are in the same direction resulting in a single large domain. The material in this state possesses maximum polarization. If the material is maintained at a high temperature (close to the transition temperature but less than the transition temperature) while the electric field is applied, the orientation of the domains is better facilitated.

The process of poling involves the following steps:

1. The material is heated to a temperature slightly less than the transition temperature and held at the temperature.
2. A sufficiently high electric field is applied to the material for about 2–3 h. All the ferroelectric domains get oriented in the direction of the electric field, and the material attains saturation polarization.
3. The material is cooled to room temperature with the electric field kept on. The domains remain frozen in the oriented state.
4. The electric field is now put off. The material remains in the maximum polarization state with most of the domains oriented in the same direction.

The poling process is illustrated in Figure 1.7.

1.8 Applications of Ferroelectric Materials

Ferroelectric materials have several device applications due to their special characteristics. As they belong to a subclass of piezoelectric and pyroelectric materials, they exhibit both piezoelectric and pyroelectric characteristics. Pyroelectric and piezoelectric characteristics in ferroelectric materials are more pronounced than in nonferroelectric materials. For most practical pyroelectric and piezoelectric devices such as infrared detectors, transducers, and actuators, ferroelectric materials are used. The various device applications of ferroelectric materials based on their special characteristics are summarized in Figure 1.8.

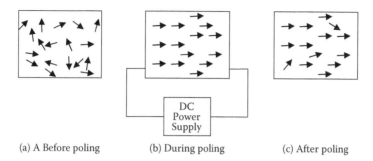

(a) A Before poling   (b) During poling   (c) After poling

FIGURE 1.7
Poling of a ferroelectric material. Each arrow represents a ferroelectric domain. (a) Unpoled state—the domains are randomly oriented. (b) High DC electric field is applied to the material—the domains get oriented in the direction of the electric field. (c) The electric field is removed—most of the domains remain aligned.

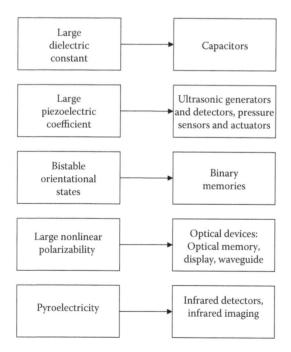

FIGURE 1.8
Applications of ferroelectric materials based on their characteristics.

2

Piezoelectric Characteristics

2.1 Introduction

Piezoelectric materials convert mechanical energy to electrical energy (direct piezoelectric effect) and electrical energy to mechanical energy (indirect piezoelectric effect).

In the direct piezoelectric effect, the input is mechanical energy and the output is electrical energy. Mechanical input can be in the form of external stress (X) or strain (x). Electrical output is in the form of surface charge density (D or P), electric field (E), or voltage (V) (Figure 2.1).

In the indirect piezoelectric effect, the input is electrical energy and the output is mechanical energy. The electrical input may be in the form of surface charge density (P/D) or electric field (E) or voltage (V), and the mechanical output is in the form of strain (x) or stress (X) on the material (Figure 2.2).

The parameters that describe the sensitivity of a piezoelectric material are the piezoelectric coefficients that relate the input and output parameters. The various piezoelectric coefficients are defined in the following section.

2.2 Piezoelectric Coefficients

In the direct piezoelectric effect, the equations that relate the mechanical input strain x to the electrical output (D/E) are

$$D = ex \tag{2.1}$$

or

$$E = hx \tag{2.2}$$

and the equations that relate the mechanical input stress X to the electrical output (D/E) are

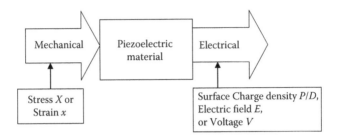

FIGURE 2.1
Direct piezoelectric effect: Input is mechanical and output is electrical.

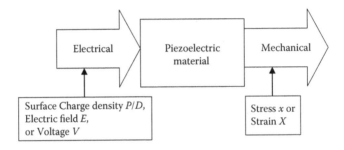

FIGURE 2.2
Indirect piezoelectric effect: Input is electrical and output is mechanical.

$$D = dX \qquad (2.3)$$

or

$$E = gX \qquad (2.4)$$

$e, h, d,$ and g are *piezoelectric coefficients* that describe the direct piezoelectric effect. Piezoelectric coefficients that describe the indirect piezoelectric effect are denoted by $e^*, d^*, h^*,$ and g^*. They are defined by the relations

$$X = e^*E \qquad (2.5)$$

$$x = d^*E \qquad (2.6)$$

$$X = h^*D \qquad (2.7)$$

$$x = g^*D \qquad (2.8)$$

Equations 2.5 and 2.6 relate the input electric field to the mechanical output (X and x), and Equations 2.7 and 2.8 relate the input charge density (D) to the mechanical output (X and x).

Using the relation between D and E (Equation 1.4), we get the following relations between the piezoelectric coefficients:

$$h = e/\varepsilon; \; h^* = e^*/\varepsilon \qquad (2.9)$$

and

$$g = d/\varepsilon; \; g^* = d^*/\varepsilon \qquad (2.10)$$

The piezoelectric effect is a transient effect, which means that the observed parameter is not an absolute value but it is the change in the parameter. In the direct piezoelectric effect, a change in strain ∂x (or in stress ∂X) causes a change in the polarization ∂D (or a change in the electric field ∂E), and in the indirect piezoelectric effect, a change in the applied filed ∂E (or polarization ∂D) causes a change in the strain ∂x (or stress ∂X). Since D and P are related by the equation $D = \varepsilon_o E + P$, the variation in D, ∂D can be replaced by the variation in P, ∂P for constant E.

The various piezoelectric coefficients defined earlier are more appropriately defined by the following partial derivatives:

Direct piezoelectric coefficients   Indirect piezoelectric coefficients

$$d = \left(\frac{\partial D}{\partial X} \right)_E \qquad\qquad d^* = \left(\frac{\partial x}{\partial E} \right)_X$$

$$g = -\left(\frac{\partial E}{\partial X} \right)_D \qquad\qquad g^* = \left(\frac{\partial x}{\partial D} \right)_X$$

$$e = \left(\frac{\partial D}{\partial x} \right)_E \qquad\qquad e^* = -\left(\frac{\partial X}{\partial E} \right)_x \qquad (2.11)$$

$$h = -\left(\frac{\partial E}{\partial x} \right)_D \qquad\qquad h^* = -\left(\frac{\partial X}{\partial D} \right)_x$$

From thermodynamics, it can be proved that

$$d = d^*, \;\; g = g^*, \;\; e = e^* \;\; and \;\; h = h^* \qquad (2.12)$$

The proof is given as follows:

The first law of thermodynamics states that an increment in the heat supplied to the system dQ results in a change dU in the internal energy of the system and causes the system to do work dW on the external surroundings:

$$dQ = dU + dW \qquad (2.13)$$

If the changes are reversible, $dQ = TdS$, where dS is the change in entropy of the system and T is the absolute temperature, so that

$$TdS = dU + dW \tag{2.14}$$

Considering the temperature T, strain x, and charge density D as independent variables, the work done on the system is $-Xdx - EdD$, so Equation 2.14 can be written as

$$TdS = dU - Xdx - EdD \tag{2.15}$$

The change in the Helmholtz energy of the system is given by

$$dA = dU - TdS - SdT \tag{2.16}$$

Substituting for $dU - TdS$ from Equation 2.15,

$$dA = Xdx + EdD - SdT \tag{2.17}$$

Since A is a perfect differential, it follows that

$$\left(\frac{\partial X}{\partial D} \right)_{x,T} = \left(\frac{\partial E}{\partial x} \right)_{D,T} \tag{2.18}$$

that is, using Equation 2.11, $h = h^*$.

Considering temperature T, stress X, and electric field E as independent variables and using the expression for Gibbs free energy $G = U - TS - Xx - ED$, we can write

$$dG = dU - TdS - SdT - Xdx - xdX - EdD - DdE$$

Using Equation 2.15,

$$dG = -SdT - xdX - DdE \tag{2.19}$$

Since G is a perfect differential, this leads to

$$\left(\frac{\partial x}{\partial E} \right)_{X,T} = \left(\frac{\partial D}{\partial X} \right)_{E,T} \tag{2.20}$$

that is, using Equation 2.11, $d = d^*$.

Using the part of the Gibbs function with only electric field applied,

$$G_1 = U - TS - ED$$

$$dG_1 = dU - TdS - SdT - EdD - DdE$$

Using Equation 2.15,

$$dG_1 = Xdx - SdT - DdE$$

which leads to

$$\left(\frac{\partial X}{\partial E} \right)_{x,T} = -\left(\frac{\partial D}{\partial x} \right)_{E,T} \tag{2.21}$$

that is, using Equation 2.11, $e = e^*$.
  Similarly, using the part of the Gibb's function with only stress applied,

$$G_2 = U - TS - Xx$$

$$dG_2 = dU - TdS - SdT - Xdx - xdX$$

Using Equation 2.15,

$$dG_2 = EdD - SdT - xdX$$

which leads to

$$\left(\frac{\partial x}{\partial D} \right)_{X,T} = -\left(\frac{\partial E}{\partial X} \right)_{D,T} \tag{2.22}$$

that is, from Equation 2.11, $g = g^*$.
  Definitions of the piezoelectric coefficients and their units are summarized in Table 2.1. Piezoelectric materials are characterized by the following parameters:

- Piezoelectric coefficients: d, g, e, h.
- Electrical parameter—permittivity ε
- Elastic parameter—compliance constant s and stiffness constant c

TABLE 2.1

Piezoelectric Coefficients—Definitions and Units

Piezoelectric Coefficient	Definition	Unit
d	$\dfrac{polarization}{stress}$	C/N
g	$\dfrac{electric\ field}{stress}$	V.m/N
e	$\dfrac{polarization}{strain}$	C/m²
h	$\dfrac{electric\ field}{strain}$	V/m
d^*	$\dfrac{strain}{electric\ field}$	m/volt
g^*	$\dfrac{strain}{polarization}$	m²/C
e^*	$\dfrac{stress}{electric\ field}$	N/V.m
h^*	$\dfrac{stress}{polarization}$	N/C

Note: It can be easily verified that the units of the pairs d,d^*; g,g^*; e,e^*; and h,h^* are the same (this can be verified by using the identity: V ≡ N.m/C).

The governing equation for the direct piezoelectric effect:

Two conditions arise:

1. Material under "free" condition; "free" refers to a state when the material is able to change dimensions with the applied field. This is the normal condition (constant stress condition).

   The governing equation is

$$D = dX + \varepsilon^X E \qquad (2.23)$$

where ε^X is the permittivity at constant stress X.

2. Material under clamped condition; "clamped" refers to either a condition where the material is physically clamped or a condition in which it is driven at a sufficiently high frequency at which the device cannot respond to the changing electric field. This is the constant strain condition.

The governing equation is

$$D = ex + \varepsilon^x E \tag{2.24}$$

where ε^x is the permittivity at constant strain x.

The governing equation for the indirect piezoelectric effect:

Two conditions arise:

1. Material short-circuited (constant electric field)

$$x = s^E X + d\,E \tag{2.25a}$$

$$X = c^E x - eE \tag{2.25b}$$

where s^E and c^E are the elastic compliance constant and the elastic stiffness constant, respectively, under constant electric field.

2. Material under open circuit (constant charge density)

$$x = s^D X + gD \tag{2.26a}$$

$$X = c^D x - hD \tag{2.26b}$$

where s^D and c^D are the elastic compliance constant and the elastic stiffness constant, respectively, under constant charge density.

2.3 Tensor Form of Piezoelectric Equations

Piezoelectric coefficients, elastic constants, and permittivity are all tensors as they relate vectors and tensors.

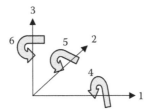

FIGURE 2.3
Right-handed Cartesian coordinate system. Directions 1, 2, and 3 represent the axes X, Y, and Z, respectively, and 4, 5, and 6 represent rotations (anticlockwise) about the three axes X, Y, and Z.

The permittivity, which relates the two vectors D and E, is a second-rank tensor. The elastic compliance constant, which relates the two second-rank tensors X and x, is a fourth-rank tensor. Piezoelectric coefficients relate stress/strain (second-rank tensors) and electric field/polarization (vectors), and so they are third-rank tensors. The system of coordinates is the right-handed Cartesian coordinate system. The X-, Y-, and Z-axes are represented by 1, 2, and 3, respectively, and the rotations about the X-, Y-, and Z-axes are represented by 4, 5, and 6 as shown in Figure 2.3.

2.3.1 Piezoelectric Coefficient *d*

Consider the direct piezoelectric effect that relates the polarization D and the stress X (Equation 2.3):

$$D = d\,X \tag{2.27}$$

Stress X is a second-rank tensor with components

$$\begin{pmatrix} X_{11} & X_{12} & X_{13} \\ X_{21} & X_{22} & X_{23} \\ X_{31} & X_{32} & X_{33} \end{pmatrix}$$

Since $X_{12} = X_{21}$, $X_{13} = X_{31}$, and $X_{23} = X_{32}$, X is symmetric with only six independent components.

D is a vector with components

$$\begin{pmatrix} D_1 \\ D_2 \\ D_3 \end{pmatrix}$$

Equation 2.27 in terms of the components becomes

$$D_1 = d_{111}X_{11} + d_{112}X_{12} + d_{113}X_{13} + d_{121}X_{21} + d_{122}X_{22} + d_{123}X_{23} + d_{131}X_{31} +$$
$$d_{132}X_{32} + d_{133}X_{33}$$

$$D_2 = d_{211}X_{11} + d_{212}X_{12} + d_{213}X_{13} + d_{221}X_{21} + d_{222}X_{22} + d_{223}X_{23} + d_{231}X_{31} +$$
$$d_{232}X_{32} + d_{233}X_{33}$$

(2.28)

$$D_3 = d_{311}X_{11} + d_{312}X_{12} + d_{313}X_{13} + d_{321}X_{21} + d_{322}X_{22} + d_{323}X_{23} + d_{331}X_{31} +$$
$$d_{332}X_{32} + d_{333}X_{33}$$

In terms of the six independent stress components,

$$D_1 = d_{111}X_{11} + d_{122}X_{22} + d_{133}X_{33} + (d_{112} + d_{121})X_{12} + (d_{123} + d_{132})X_{23} +$$
$$(d_{113} + d_{131})X_{13}$$

$$D_2 = d_{211}X_{11} + d_{222}X_{22} + d_{233}X_{33} + (d_{212} + d_{221})X_{12} + (d_{223} + d_{232})X_{23} +$$
$$(d_{213} + d_{231})X_{13}$$

(2.29)

$$D_3 = d_{311}X_{11} + d_{322}X_{22} + d_{333}X_{33} + (d_{312} + d_{321})X_{12} + (d_{323} + d_{332})X_{23} +$$
$$(d_{313} + d_{331})X_{13}$$

For simplicity, *reduced matrix notation* is used for the second-rank tensor X_{ij}. The two indices i and j, each taking values 1 to 3, are replaced by a single index that takes values 1 to 6.
That is,

$$11 \equiv 1; 22 \equiv 2; 33 \equiv 3; 23 = 32 \equiv 4; 31 = 13 \equiv 5; 12 = 21 \equiv 6$$

The subscripts 1, 2, and 3 indicate tensile or compressive stress (or strain) and 4, 5, and 6 indicate shear stress (or shear strain)—rotation about axis 1, 2, and 3, respectively, (Figure 2.3).
Thus,

$$X_{11} \equiv X_1; \quad X_{22} \equiv X_2; \quad X_{33} \equiv X_3; \quad X_{23} = X_{32} \equiv X_4;$$
$$X_{31} = X_{13} \equiv X_5; \quad X_{12} = X_{21} \equiv X_6$$

So the first of Equation 2.29 is written as

$$D_1 = d_{11}X_1 + d_{12}X_2 + d_{13}X_3 + d_{14}X_4 + d_{15}X_5 + d_{16}X_6 \qquad (2.30)$$

where

$$d_{11} \equiv d_{111}; \quad d_{12} \equiv d_{122}; \quad d_{13} \equiv d_{133}; \quad d_{14} \equiv d_{123} + d_{132};$$
$$d_{15} \equiv d_{113} + d_{131}; \quad d_{16} \equiv d_{112} + d_{121}$$

and

$$d_{123} = d_{132} = d_{14}/2$$

$$d_{113} = d_{132} = d_{15}/2$$

$$d_{112} = d_{121} = d_{16}/2$$

Similarly, the other two equations are written as

$$D_2 = d_{21}X_1 + d_{22}X_2 + d_{23}X_3 + d_{24}X_4 + d_{25}X_5 + d_{26}X_6 \qquad (2.31)$$

$$D_3 = d_{31}X_1 + d_{32}X_2 + d_{33}X_3 + d_{34}X_4 + d_{35}X_5 + d_{36}X_6 \qquad (2.32)$$

Equations 2.30–2.32 may be written in the matrix form:

$$\begin{pmatrix} D_1 \\ D_2 \\ D_3 \end{pmatrix} = \begin{pmatrix} d_{11} & d_{12} & d_{13} & d_{14} & d_{15} & d_{16} \\ d_{21} & d_{22} & d_{23} & d_{24} & d_{25} & d_{26} \\ d_{31} & d_{32} & d_{33} & d_{34} & d_{35} & d_{36} \end{pmatrix} \begin{pmatrix} X_1 \\ X_2 \\ X_3 \\ X_4 \\ X_5 \\ X_6 \end{pmatrix} \qquad (2.33)$$

Thus, the piezoelectric coefficient d is represented as a 3×6 matrix.

If the stress matrix [X] on the right-hand side of Equation 2.33 is replaced by the strain matrix [x], then the piezoelectric coefficient matrix is replaced by the [e] matrix (see Equation 2.1).

Similarly, by replacing the vector [D] on the left-hand side by the vector [E], the piezoelectric coefficient matrices [g] and [h] (of Equations 2.4 and 2.2) can be represented as 3×6 matrices.

In the indirect piezoelectric effect, the equation that relates the electric field E with the stress X (Equation 2.5) is written in the matrix form as

$$\begin{pmatrix} X_1 \\ X_2 \\ X_3 \\ X_4 \\ X_5 \\ X_6 \end{pmatrix} = \begin{pmatrix} e_{11} & e_{12} & e_{13} \\ e_{21} & e_{22} & e_{23} \\ e_{31} & e_{32} & e_{33} \\ e_{41} & e_{42} & e_{43} \\ e_{51} & e_{52} & e_{53} \\ e_{61} & e_{62} & e_{63} \end{pmatrix} \begin{pmatrix} E_1 \\ E_2 \\ E_3 \end{pmatrix} \tag{2.34}$$

Thus, in the indirect piezoelectric effect, the piezoelectric coefficient e is represented as a 6×3 matrix.

If the stress matrix on the left-hand side of Equation 2.34 is replaced by the strain matrix $[x]$, then the piezoelectric coefficient matrix is replaced by the $[d]$ matrix (see Equation 2.6). Similarly, by replacing the vector $[E]$ on the right-hand side of Equation 2.34 by D, the piezoelectric coefficient matrices $[g]$, $[h]$ (of Equations 2.7 and 2.8) can be represented as 6×3 matrices.

2.3.2 Mechanical Parameters: Elastic Compliance Constant and Elastic Stiffness Constant

The main mechanical parameters of interest in piezoelectric materials are elastic compliance constant and elastic stiffness constants. These two constants relate the two second-rank tensors stress and strain, and so they are fourth-rank tensors.

Elastic compliance constant s is defined by the relation

$$x_{ij} = s_{ijkl} X_{kl}$$

Elastic stiffness constant c is defined by the relation

$$X_{ij} = c_{ijkl} x_{kl}$$

The subscripts i, j, k, and l each takes values 1, 2, and 3. The equations in the reduced matrix notation defined earlier would be written as

$$x_i = s_{ij} X_j \tag{2.35}$$

and

$$X_i = c_{ij} x_j \tag{2.36}$$

where the subscripts i and j each take values 1 to 6.

The expanded form of Equation 2.35:

$$\begin{bmatrix} x_1 \\ x_2 \\ x_3 \\ x_4 \\ x_5 \\ x_6 \end{bmatrix} = \begin{bmatrix} s_{11} & s_{12} & s_{13} & s_{14} & s_{15} & s_{16} \\ s_{21} & s_{22} & s_{23} & s_{24} & s_{25} & s_{26} \\ s_{31} & s_{32} & s_{33} & s_{34} & s_{35} & s_{36} \\ s_{41} & s_{42} & s_{43} & s_{44} & s_{45} & s_{46} \\ s_{51} & s_{52} & s_{53} & s_{54} & s_{55} & s_{56} \\ s_{61} & s_{62} & s_{63} & s_{64} & s_{65} & s_{66} \end{bmatrix} \begin{bmatrix} X_1 \\ X_2 \\ X_3 \\ X_4 \\ X_5 \\ X_6 \end{bmatrix} \tag{2.37}$$

An identical matrix equation can be written for Equation 2.36. Thus, the compliance and stiffness constants are represented as 6 × 6 matrices. The two matrices are symmetric; that is, the components on the right side of the diagonal have equal components symmetrically on the left side of the diagonal. So, of the 36 components, only 21 components are independent.

2.3.3 Dielectric Parameter: Permittivity

The dielectric parameter of interest in piezoelectric materials is the permittivity ε, which relates the vectors D and E, and so it is a second-rank tensor:

$$D_i = \varepsilon_{ij} E_j \tag{2.38}$$

The subscripts i and j each take values 1, 2, and 3. The tensor form of the equation is

$$\begin{bmatrix} D_1 \\ D_2 \\ D_3 \end{bmatrix} = \begin{bmatrix} \varepsilon_{11} & \varepsilon_{12} & \varepsilon_{13} \\ \varepsilon_{21} & \varepsilon_{22} & \varepsilon_{23} \\ \varepsilon_{31} & \varepsilon_{32} & \varepsilon_{33} \end{bmatrix} \begin{bmatrix} E_1 \\ E_2 \\ E_3 \end{bmatrix} \tag{2.39}$$

Thus, the permittivity is a 3 × 3 tensor.

2.4 Independent Components of Piezoelectric, Elastic, and Dielectric Matrices

Depending on the symmetry of the piezoelectric system, the components of the tensors defined earlier are all not independent. So, the number of components in each of these tensors gets reduced.

For example, most of the practical piezoelectric materials used in sensor and actuator applications are polycrystalline or noncrystalline. Thus, they are isotropic. But because they are poled in a specific direction, the system loses its isotropic nature. Normally, the poling direction is taken as the z-direction (direction 3). The symmetry of such a system with one unique direction is described in crystallographic notation as ∞ mm. This means that the system has ∞ fold symmetry axis along the poled direction and an infinite set of reflection planes parallel to the symmetry axis. The system is said to be orthotropic or cylindrically symmetric. The number of independent components of the piezoelectric, elastic, and dielectric matrices for such a system gets enormously reduced. The piezoelectric coefficient matrices ([d], [e], [h], [g]) will have only three independent components; the elastic constant matrices [s] and [c] will have four independent components, and the dielectric permittivity will have two independent components. The reduced matrices are as follows:

The piezoelectric coefficient matrix:

$$[d] = \begin{bmatrix} 0 & 0 & 0 & 0 & d_{15} & 0 \\ 0 & 0 & 0 & d_{15} & 0 & 0 \\ d_{31} & d_{31} & d_{33} & 0 & 0 & 0 \end{bmatrix} \tag{2.40}$$

or

$$[d] = \begin{bmatrix} 0 & 0 & 0 \\ 0 & 0 & 0 \\ d_{31} & d_{31} & d_{33} \\ 0 & d_{15} & 0 \\ d_{15} & 0 & 0 \\ 0 & 0 & 0 \end{bmatrix} \tag{2.41}$$

with $d_{13} = d_{31}$ and $d_{15} = d_{51}$.

Similar matrices can be written for other piezoelectric coefficients [e], [g], [h].

The elastic compliance constant matrix:

$$[s] = \begin{bmatrix} s_{11} & s_{12} & s_{13} & 0 & 0 & 0 \\ s_{12} & s_{11} & s_{13} & 0 & 0 & 0 \\ s_{13} & s_{13} & s_{33} & 0 & 0 & 0 \\ 0 & 0 & 0 & s_{44} & 0 & 0 \\ 0 & 0 & 0 & 0 & s_{44} & 0 \\ 0 & 0 & 0 & 0 & 0 & 2(s_{11} - s_{12}) \end{bmatrix} \tag{2.42}$$

with $s_{11} = s_{22}$; $s_{12} = s_{21}$; $s_{13} = s_{31}$; $s_{44} = s_{55}$; and $s_{66} = 2(s_{11} - s_{12})$.

The dielectric constant matrix:

$$[\varepsilon] = \begin{bmatrix} \varepsilon_1 & 0 & 0 \\ 0 & \varepsilon_1 & 0 \\ 0 & 0 & \varepsilon_3 \end{bmatrix} \tag{2.43}$$

with $\varepsilon_1 = \varepsilon_2$.

However, it should be remembered that for a single piezoelectric crystal sample the number of independent components will depend on the symmetry of the crystal and will be more than what it is for the poled polycrystalline samples.

2.4.1 Interpretation of Piezoelectric Coefficients d_{33}, d_{31}, and d_{15}

d_{33}: The coefficient relates D and X by the relation

$$D_3 = d_{33}X_3$$

The equation in conventional notation would be

$$D_3 = d_{333}X_{33}$$

The equation describes the direct piezoelectric effect in which the external stimulus is the tensile or compressive stress X_{33} along direction 3 on face 3 and the response is the charge density D_3 developed on face 3 as illustrated in Figure 2.4a. This is called the *longitudinal mode*.

d_{31}: The coefficient relates D and X by the relation

$$D_3 = d_{31}X_1$$

The equation in conventional notation would be

$$D_3 = d_{311}X_{11}$$

The equation describes the direct piezoelectric effect in which the external stimulus is the tensile or compressive stress in direction 1 on face 1 and the response is the charge density on face 3 as illustrated in Figure 2.4b. This is called the *transverse mode*.

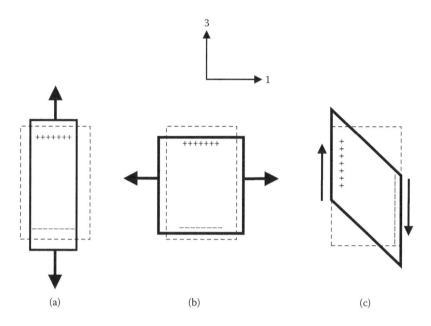

FIGURE 2.4
Direct piezoelectric effect. (a) Longitudinal mode—d_{33}: Tensile stress direction 3 resulting in charges on face 3 (b) Transverse mode—d_{31}: Tensile stress on face 1 resulting in charges on face 3. (c) Shear mode—d_{15}: Shear stress on face 1 in direction 3 (X_{13}) resulting in charges on face 1.

d_{15}: The coefficient relates D and X by the relation

$$D_1 = d_{15}X_5$$

The equation in the conventional notation would be

$$D_1 = d_{113}X_{13}$$

The equation describes the direct piezoelectric effect in which the external stimulus is the shear stress X_{13}, that is, parallel stress on face 1 (in direction 3), and the response is the charge density on face 1 as illustrated in Figure 2.4c. This is called the *shear mode*.

Similarly, the coefficients e_{33}, e_{31}, and e_{15} can be interpreted in the indirect piezoelectric effect.

e_{33}:

$$X_3 = e_{33}E_3 \ \text{ or } \ X_{33} = e_{333}E_3$$

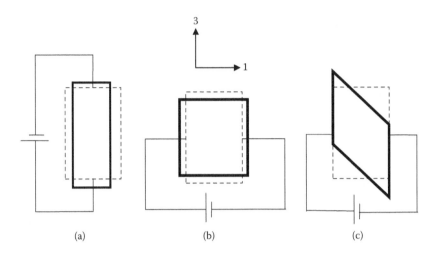

FIGURE 2.5
Indirect piezoelectric effect. (a) Longitudinal mode—e_{33}: Electric field applied in direction 3 resulting in tensile strain in direction 3. (b) Transverse mode—e_{31}: Electric field applied in direction 1 resulting in compressive strain in direction 3. (c) Shear mode—e_{15}: Electric field applied in direction 1 resulting shear strain (x_{13}) on face 1.

The equation describes the indirect piezoelectric effect in which the external stimulus is the electric field E_3 applied in direction 3 and the response is the tensile or compressive stress experienced by the material in direction 3 on face 3. This is the longitudinal mode illustrated in Figure 2.5a.

e_{31}:

$$X_3 = e_{31}E_1 \ \text{ or } \ X_{33} = e_{331}E_1$$

The equation describes the indirect piezoelectric effect in which the external stimulus is the electric field E_1 applied in direction 1 and the response is the tensile or compressive stress experienced by the material in direction 3 on face 3. This is illustrated in Figure 2.5b.

e_{51}:

$$X_5 = e_{51}E_1 \ \text{ or } \ X_{31} = e_{311}E_1$$

The equation describes the indirect piezoelectric effect in which the external stimulus is the electric field E_1 applied in direction 1 and the response is the shear stress experienced by the material on face 1. This is illustrated in Figure 2.5c.

2.5 Generator and Motor Actions of Piezoelectric Materials

2.5.1 Generator Action

The direct piezoelectric effect is called the *generator action* because mechanical energy gets converted to electric field. The charges developed on the faces of the material generate an electric field that can be measured as an external voltage.

The generator actions in longitudinal mode, transverse mode, and shear mode are illustrated in Figure 2.6a,b,c.

In the longitudinal mode, the electric field generated is related to the stress through the coefficient g_{33} by (Equation 2.4)

$$E_3 = g_{33}X_3$$

If t is the thickness of the sample and l and w are the length and width, respectively (see Figure 2.6a), then the voltage developed on applying an external mechanical force F can be expressed in terms of the dimensions of the sample as

$$\frac{V}{t} = g_{33}\frac{F_3}{lw}$$

In the transverse mode (Figure 2.6b), the relation is

$$E_3 = g_{31}X_1$$

In terms of the dimensions of the sample,

$$\frac{V}{t} = g_{31}\frac{F_1}{tw}$$

The voltage generated is directly proportional to the external force applied.

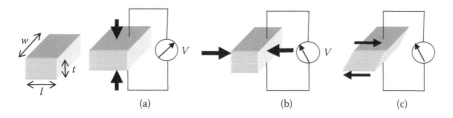

(a)   (b)   (c)

FIGURE 2.6
Generator action of piezoelectric material: (a) longitudinal mode, (b) transverse mode, (c) shear mode.

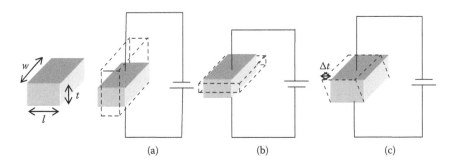

FIGURE 2.7
Motor action of a piezoelectric material: (a) tensile mode, (b) compressive mode, (c) shear mode.

2.5.2 Motor Action

The indirect piezoelectric effect is called the *motor action* because electrical energy gets converted to mechanical energy. Application of the electric field generates strain in the material. The motor actions in the longitudinal, transverse, and shear modes are illustrated in Figure 2.7a,b,c.

In the longitudinal mode, the strain and the electric field are related by Equation 2.6:

$$x_3 = d_{33}E_3$$

In terms of the voltage and sample dimensions, we can write

$$\frac{\Delta t}{t} = d_{33}\frac{V}{t}$$

or

$$\Delta t = d_{33}V$$

The change in dimension Δt is directly proportional to the applied voltage.
In the shear mode (Figure 2.6c), the relation is

$$\Delta t = d_{15}V$$

where Δt denotes the shear strain.

2.6 Strain versus Electric Field in Piezoelectric Materials

A typical strain versus electric field curve of a piezoelectric material, which is also ferroelectric (PZT), is shown in Figure 2.8. The strain in longitudinal

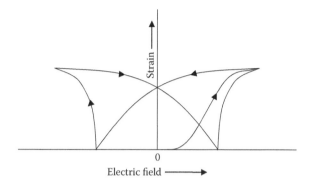

FIGURE 2.8
Strain versus electric field for a typical piezoelectric material.

and transverse modes both exhibit hysteresis effects. The strain increases with the applied electric field initially linearly and later gets saturated. The strain is of the order of about 10^{-3} for an electric field in the range 1–1.5 kV for PZT, which is the best-known piezoelectric ceramic, used widely for transducer applications. For other piezoelectric materials, it is much less. When the applied electric field is gradually reduced, the strain does not follow the same curve. The change in strain lags behind the change in electric field. When the electric field is zero, a remnant strain is seen in the material. If the electric field is increased in the reverse direction, the strain becomes zero at a particular negative electric field. If the electric field is now increased in the same direction, the strain increases and attains saturation. Thus, the strain versus electric field is a symmetric curve as shown in Figure 2.8. A closed loop is formed when the electric field in the reverse direction is again brought to zero and increased in the positive direction as shown Figure 2.8.

2.7 Piezoelectric Coupling Coefficient k

Piezoelectric coupling coefficient is a measure of the efficiency of a piezoelectric material as a transducer. It quantifies the ability of the piezoelectric material to convert one form of energy (mechanical or electrical) to the other form (electrical or mechanical). It is defined by

$$k^2 = \frac{(Piezoelectric\ energy\ density\ stored\ in\ the\ material)^2}{Electrical\ energy\ density \times Mechanical\ energy\ density} \tag{2.44}$$

If the electrical energy density is W_e, and the mechanical energy density is W_m, then k^2 is written as

$$k^2 = \frac{(W_{em})^2}{W_e W_m} \tag{2.45}$$

where W_{em} is the piezoelectric energy density.

The expression for k^2 can be obtained in terms of the piezoelectric coefficients:

From Equations 2.23 and 2.25, we have

$$D = dX + \varepsilon^X E$$

$$x = s^E X + d E$$

Multiplying the first equation by ½ E and the second equation by ½ X, we get

$$\frac{1}{2} DE = d\frac{1}{2}EX + \varepsilon^X \frac{1}{2}E^2 = W_{em} + W_e \tag{2.46}$$

$$\frac{1}{2} xX = \frac{1}{2}s^E X^2 + d\frac{1}{2}EX = W_m + W_{em} \tag{2.47}$$

where

$$W_{em} = d\frac{1}{2}EX \quad \text{is the piezoelectric energy}$$

$$W_m = \frac{1}{2}s^E X^2 \quad \text{is the mechanical energy}$$

and

$$W_e = \frac{1}{2}\varepsilon^X E^2 \quad \text{is the electrical energy}$$

Substituting expressions for energy in Equation 2.45, k^2 is obtained in terms of piezoelectric constants as

$$k^2 = \frac{d^2}{\varepsilon^X s^E} \tag{2.48}$$

The coupling coefficient is the ratio of useable energy delivered by the piezoelectric element to the total energy taken up by the element. The piezoelectric element manufacturers normally specify the theoretical k values

TABLE 2.2

Piezoelectric Coupling Coefficients—Definitions

Notation	Definition
k_{33}	Coupling coefficient when electric field is in direction 3 and mechanical vibrations are in direction 3.
k_{31}	Coupling coefficient when electric field is in direction 3 and mechanical vibrations are in direction 1.
k_t	Coupling coefficient for a thin disc in which electric field is in direction 3 (across the thickness of the disc along which the disc is poled) and mechanical vibrations are in the same direction (direction 3)
k_p	Coupling coefficient for a thin disc in which electrical field is in direction 3 (across the thickness of the disc along which it is poled) and mechanical vibrations are along radial direction.

that can be in the range of 30%–75%. In practice, k values depend on the design of the device and the directions of the applied stimulus and the measured response.

The coupling coefficient is written with subscripts just like piezoelectric constants to denote the direction of the external stimulus and the direction of measurement. The various coupling coefficients are defined in Table 2.2.

2.8 Dynamic Behaviour of a Piezoelectric Material

When a piezoelectric material is subjected to an AC electric field, the dimensions of the material change periodically; in other words, the material experiences vibration at the same frequency as that of the applied field. In the direct effect, when a vibrating force is applied to the piezoelectric material, it generates an oscillating electric field at the same frequency.

For the analysis of the dynamic behaviour of a vibrating piezoelectric material, an equivalent electrical circuit is used, drawing analogy between mechanical and electrical components. This is illustrated in Figure 2.9. The vibrating force applied to the material is analogous to an alternating voltage. The piezoelectric element behaves as a capacitor of capacitance $C_o = \varepsilon A/d$, where ε is the permittivity of the material and A and d are the area and thickness of the element, respectively. The mass M (inertia) of the piezoelectric element is equivalent to the inductance L, and the compliance constant is equivalent to the capacitor C. The energy loss due to friction is equivalent to the energy loss due to electrical resistance r in the circuit.

The impedance of the vibrating system is a function of frequency. The variation of the impedance as a function of frequency for such a system is shown in Figure 2.10. The impedance shows a minimum and a maximum as

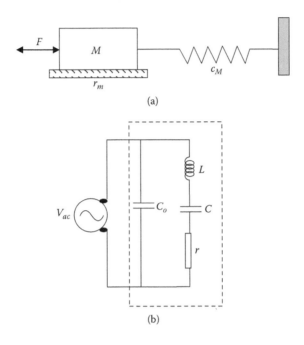

(a)

(b)

FIGURE 2.9
Vibrating piezoelectric material can be represented by a mechanical system or by an equivalent electrical circuit. (a) Mechanical system: F is the vibrating force applied on the system of mass M (inertia). r_m represents the mechanical resistance due to friction that causes energy loss. c_M represents the spring constant of the mechanical system. (b) Equivalent circuit: V, applied AC voltage equivalent to the force F. L, inductance equivalent to the mass M. r, electrical resistance equivalent to mechanical friction (energy loss). C, capacitance equivalent to the compliance (related to the spring constant) of the material. C_o, the electrical capacitance of the piezoelectric material.

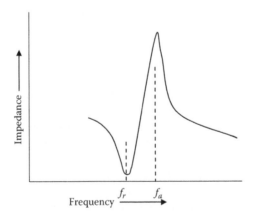

FIGURE 2.10
Frequency response of a piezoelectric element. The impedance shows a minimum at f_r and a maximum at f_a. f_r is called the *resonance frequency*, and f_a is called the *antiresonance frequency*.

shown in the figure. The frequency at which the impedance is a minimum is called the *resonance frequency*, and the frequency at which the impedance is a maximum is called the *antiresonance frequency*.

At the resonance frequency f_m, the piezoelectric system vibrates with maximum amplitude. The resonance frequency f_m is equal to the series resonance frequency f_s at which the impedance of the equivalent electrical circuit is zero, assuming that the resistance due to mechanical loss is zero:

$$f_m = f_s = \frac{1}{2\pi}\sqrt{\frac{1}{LC}}$$

The antiresonance frequency f_a is equal to the parallel resonance frequency f_p of the equivalent electrical circuit, assuming that the resistance due to mechanical loss is zero.

$$f_a = f_p = \frac{1}{2\pi}\sqrt{\frac{C+C_o}{LCC_o}}$$

Resonance and antiresonance frequencies can be experimentally measured for a piezoelectric element. The values of f_r and f_a can be used to evaluate the electromechanical coupling coefficient k. The relation between coupling coefficient and the frequencies f_r and f_a depends on the shape of the piezoelectric element.

3

Piezoelectric Materials

3.1 Introduction

Classification of dielectric materials into different classes as piezoelectric, pyroelectric, and ferroelectric is described in Chapter 1. This is briefly repeated here for clarity and relevance. Piezoelectric materials are a special class of dielectric materials that exhibit transducer characteristics by virtue of their noncentrosymmetric crystal structure. Some piezoelectric materials possess spontaneous polarization that decreases with increasing temperature, and these are classified as pyroelectric materials. In some of these pyroelectric materials, the spontaneous polarization can be reoriented by applying an external electric field, and such materials are classified as ferroelectric materials. Thus, ferroelectric materials possess both pyroelectric and piezoelectric characteristics.

Ferroelectric materials exhibit better piezoelectric and pyroelectric characteristics than nonferroelectric materials. Most piezoelectric devices are based on ferroelectric materials as their transducer, and actuator characteristics are better than those of nonferroelectric piezoelectric materials. Quartz is an example of a nonferroelctric piezoelectric material that is widely used not as an actuator or transducer, but as a resonator and a sensor. A single crystal of quartz is highly suited for resonator applications because of its robust mechanical characteristics.

This chapter describes the fabrication and characteristics of piezoelectric materials that are widely used in engineering and medical applications.

3.2 Quartz

Quartz is the crystalline form of silicon dioxide, and it is a naturally occurring piezoelectric material. It is the most widely used nonferroelectric piezoelectric material. Because of its piezoelectric property, quartz crystal in the form of a thin plate or a tuning fork can be mechanically excited using an electrical signal. An AC signal to the quartz plate or tuning fork makes it vibrate (indirect piezoelectric effect), and if the frequency of the AC signal matches the natural frequency of the crystal, it vibrates with maximum amplitude. As the quartz crystal vibrates, it produces an AC signal of the same frequency (direct piezoelectric effect). This characteristic is used in a feedback system to get an extremely stable frequency output. In principle, any piezoelectric material can be used for frequency control applications, but quartz crystal has special advantages over other piezoelectric materials such as robust mechanical properties, high stiffness constant, high Q factor, good reliability, and long life. Another characteristic that makes it highly suitable is that its behaviour is unaffected by temperature and other environmental changes.

Naturally occurring or synthetic quartz (SiO_2) is used for the growth of single crystals. A solution growth technique called *hydrothermal growth* is used. An alkali metal hydroxide or carbonate solution of quartz is taken in a large steel vessel (autoclave), and seed crystals are hung on top of the vessel. A precisely controlled temperature gradient is maintained across the length of the steel vessel, and a high pressure is generated in the vessel. The temperature is in the range of 380°C to 400°C, and the pressure is in the range of 15,000 to 30,000 psi. The time taken for growing large size crystals is about several weeks.

A single crystal of quartz is shown in Figure 3.1a. Thin plates of quartz are cut at specific crystallographic directions for applications in resonators. The angle of cut with respect to the crystallographic directions determines the device parameters. The commonly used "crystal cut" for resonator applications is called the "AT" cut, which is shown in Figure 3.1b. The cut plane makes an angle 35.25° with the crystallographic z-direction. In this orientation, the crystal exhibits high frequency stability with respect to the change in temperature. For crystal oscillators, quartz crystals are used in the form of thin plates or in the form of tuning forks. The chemical etching technique

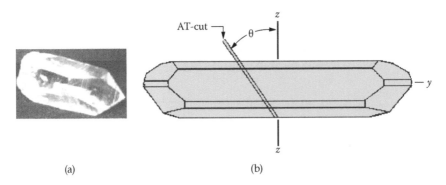

(a)   (b)

FIGURE 3.1
(a) Single crystal of quartz, (b) AT-cut in a quartz crystal. θ = 35.25°.

is used to form tiny tuning forks of quartz. AT-cut quartz crystals are available in the range of 30 kHz up to 250 MHz and above.

Many different types of cuts other than AT-cut are used for various applications. Quartz plates are mostly used in thickness-mode vibrations, although other modes such as shear-mode and extensional-mode vibrations are useful in some applications. In the thickness-mode vibrations, the thinner the plate the higher will be the fundamental frequency. Practical difficulties in making very thin plates limit the highest fundamental frequency that can be obtained to 45 MHz. For applications at higher frequencies, higher-mode vibrations (overtones) that are multiples of the fundamental modes are used.

Quartz crystal oscillators are used wherever an extremely precise frequency control is required. Some of the applications are electronic clocks and watches, clock signal generators in computers, and microprocessor-based instruments, transmitters, and receivers for communication.

For low-frequency applications such as clocks and watches, quartz crystals in the form of tuning forks are used. The frequency normally selected for the application is 32.768 kHz. For other applications such as transmitters/receivers, computer clocks, and atomic force microscope, quartz crystals of much higher frequencies, of the order of several megahertz, are used.

Applications of quartz crystal as sensors (pressure sensor, gyroscope, and microphone) and as resonators in crystal oscillator, microbalance, and atomic microscope are described in Chapter 4.

3.3 Lead Zirconate Titanate (PZT)

3.3.1 Composition and Structure of PZT

Lead zirconate titanate (PZT) is a ferroelectric ceramic that is most widely used in most of the transducer and actuator applications. It is one of the best-known piezoelectric materials, with excellent piezoelectric and mechanical properties. Another advantage of the material is that it can be used for transducer applications in polycrystalline form. It is prepared in the form of fine powder and can easily be formed into any required shape: disc, cylinder, plate, or thin film.

The chemical formula of lead zirconate titanate is $PbZr(Ti)O_3$. It is a solid solution of two oxides: lead zirconate ($PbZrO_3$), which is antiferroelectric, and lead titanate ($PbTiO_3$), which is ferroelectric. The crystal structure of PZT is perovskite. The unit cell is shown in Figure 3.2. The unit cell contains one molecule with the lead atoms at the corners (1 atom), oxygen atoms at the face centres (3 atoms), and Zr or Ti atoms at the body centre. The cubic structure exists at temperatures greater than 350°C at which the material is paraelectric. The zirconium and titanium atoms are octahedrally coordinated to the

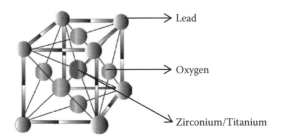

FIGURE 3.2
Unit cell of PZT. The unit cell contains one molecule. Pb at cube corners, oxygen at face centres, and Zr or Ti atom at the body centre.

oxygen atoms. The ionic radii of lead and oxygen ions are both equal and approximately equal to 0.14 nm. The lattice parameter of the cubic structure is about 0.40 nm.

3.3.2 Phase Transitions in PZT

The structure of PZT is cubic at high temperature, that is, above the Curie temperature (which is about 350°C). On cooling, the structure undergoes a displacive phase transformation with atomic displacements of about 0.1 Å. The phase diagram is shown in Figure 3.3.

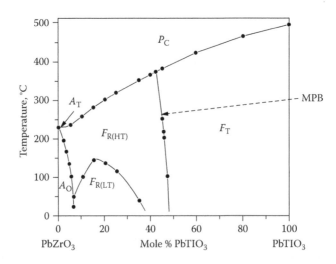

FIGURE 3.3
Phase diagram of PZT. P_C—paraelectric cubic phase. F_T—ferroelectric tetragonal phase. $F_{R(HT)}$—ferroelectric high-temperature rhombohedral phase. $F_{R(LT)}$—ferroelectric low-temperature rhombohedral phase. A_o—antiferroelectric phase. (Courtesy: L. E. Cross, 1993, *Ferroelectric Ceramics-Tutorial Reviews, Theory, Processing and Applications*, N. Setter and E. L. Colla, ed., 1, Birkhauser Verlag, Basel.) The morphotropic phase boundary (MPB) is indicated by the arrow.

Above the Curie temperature, the structure is cubic, and the material is in the paraelectric phase (P_C). The Curie temperature itself varies with the composition as seen in the phase diagram. The crystal structure below the Curie temperature depends on the composition of the solid solution. In titanium-rich compositions (lead titanate greater than 50%), on cooling from above the Curie temperature, the material becomes ferroelectric at the Curie temperature, and the structure changes from cubic to tetragonal. The tetragonal structure persists up to 0 K (F_T). In zirconium-rich compositions (lead zirconate greater than 50%), on cooling, the material becomes ferroelectric at the Curie temperature, and the structure changes from cubic to rhombohedral. There are two forms of rhombohedral structures: one form exists at higher temperatures (R_{HT}) and the other at lower temperatures (R_{LT}) as shown in the figure. At very low concentrations of lead titanate below 5%, the solid solution is antiferroelectric below the Curie temperature as shown in the figure.

Although below the Curie temperature PZT of all compositions above 5% of lead titanate is ferroelectric, the solid solution is useful as a practical piezoelectric ceramic only over a small composition of about 48% of $PbTiO_3$ and 52% of $PbZrO_3$. This composition lies near the boundary of the two phases, rhombohedral ($F_{R(HT)}$) and tetragonal (F_T), as shown in the phase diagram. This boundary is called the *morphotropic phase boundary* (MPB). At this composition, PZT is poised between the two structures, rhombohedral and tetragonal, with equal probability. Both phases are ferroelectric, the easy directions of polarization in the tetrahedral phase being <100> directions (six possible orientations) and in the rhombohedral phase being <111> directions (eight possible orientations). Thus, when the two structures coexist at the MPB, there are totally 14 easy directions of polarizations. This makes the material most sensitive as a piezoelectric material at this composition. The ferroelectric nature at this composition exists in the temperature range from near 0°C up to about 350°C (Curie temperature). Figure 3.4 shows the piezoelectric coefficient d_{33} and dielectric constant of PZT as a function of composition. Both the d_{33} coefficient and the dielectric constant show a peak at the composition of 48% $PbTiO_3$ and 52% $PbZrO_3$. Most of the commercially available PZT transducer materials have compositions close to this value.

3.3.3 Lanthanum-Doped PZT (PLZT)

The piezoelectric properties of PZT have been studied widely, and the focus of researchers has been on improving the properties by addition of donor or acceptor impurities to PZT. Substituting Pb sites with trivalent (donor) or monovalent (acceptor) impurities affects the dielectric and piezoelectric properties because of the vacancies generated. Among the various dopants, the donor impurity La^{3+} has been studied the most as it is found to improve the piezoelectric and dielectric properties of PZT [1–3]. Commercially available PZT materials are mostly lanthanum doped (termed PLZT), and they have a higher piezoelectric coefficient and dielectric constant values than undoped PZT.

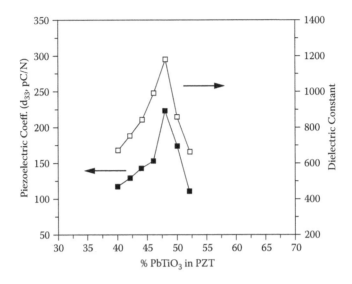

FIGURE 3.4
Piezoelectric coefficient d_{33} and dielectric constant as a function of composition. (Courtesy: B. Jaffe, W. R. Cook, Jr., and H. Jaffe, 1971, *Piezoelectric Ceramics*, Academic Press, London.)

3.4 Fabrication of PZT

3.4.1 Fabrication of PZT in Powder Form

Several techniques have been adopted for fabrication of PZT [4–9]. The most commonly used techniques are

- Solid-state reaction technique
- Coprecipitation technique
- Sol–gel technique

PZT in the form of fine powders can be obtained from the previous techniques. Other methods such as tape-casting and chemical vapour deposition techniques are used for obtaining PZT in the form of thick or thin films.

3.4.1.1 Solid-State Solution Technique

In this technique, the oxides (PbO, TiO_2, and ZrO_2) in suitable proportions are mixed well and subjected to solid-state reaction by the calcination process. The calcination process involves heating the oxides to about 650°C and maintaining at that temperature for about 2 to 3 h. The product is then heated to about 850°C. The mixture is milled to obtain a particle size of about 1 μm. Ball milling is done using zirconia balls to avoid contamination during milling. The process has been standardized and optimized to get submicronsized powder with a very narrow particle size distribution.

3.4.1.2 Coprecipitation from Solution

The coprecipitation technique is a wet chemical technique in which an aqueous solution of the oxides containing a precipitating agent is first prepared. The precipitated product is obtained by filtration. The filtered product is then dried and subjected to a thermal process for the formation of a desired compound. The important parameters to be controlled are the pH of the solution, mixing rates, the proportion of the oxides, and the temperature. PZT powder of high purity and fine particles can be obtained by this technique.

3.4.1.3 Sol–Gel Technique

The sol–gel technique consists of preparation of a solution of components that are in the form of soluble precursor compounds (metal–organic precursors). The polymerization reaction takes place, giving rise to three-dimensional structures and gel formation. The gel is then dried and ground to get fine powders of PZT. The powders synthesized by this method have a high density close to that of the theoretical value and are of highly pure quality. The advantage of the method is that it enables synthesis at a relatively low temperature, which reduces the vaporization of components, especially PbO. This ensures the maintenance of the stoichiometric ratio of the product. It is not commonly used for bulk preparation, because the initial raw materials required are expensive.

3.4.2 Shape Forming Techniques from PZT Powder

3.4.2.1 PZT Discs, Cylinders, Plates, etc.

For transducer and actuator applications, PZT is required in the form of discs, cylinders, or plates of different dimensions. The PZT powders fabricated by the different techniques described earlier are used to form products of desired shapes and sizes. The powder is initially mixed with a polymer binder and pressed in moulds using high pressure. The techniques used for pressing are

- Uniaxial pressing
- Isostatic pressing

In uniaxial pressing, the powder is compacted in a rigid die by applying pressure along a single axis using pistons (Figure 3.5a). In isostatic pressing,

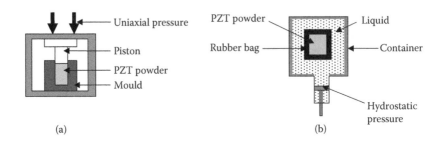

FIGURE 3.5
PZT shape forming: (a) uniaxial pressure, (b) isostatic pressure.

the pressure is applied uniformly from all sides. This method gives better uniformity of green density than uniaxial pressing. Isostatic pressing is achieved by keeping the powder in a rubber bag and immersing the bag in a liquid which acts as a pressure transmitter. Hydrostatic pressure is applied on the rubber bag to compact the powder (Figure 3.5b).

3.4.2.2 PZT Thick Films—Tape Casting Technique

The tape casting technique is a commonly used technique to obtain PZT in the form of thick films. The arrangement is shown in Figure 3.6. A fine powder of PZT is suspended in aqueous or nonaqueous liquid systems consisting of solvents and binders to form a slurry. The slurry is allowed to flow onto a cellulose sheet spread on a carrier. The carrier is passed under a doctor blade as shown in the figure. A green sheet of uniform thickness is formed on the cellulose sheet. The solvent is then allowed to evaporate, leaving dense flexible PZT thick film that can be stripped from the sheet. Large-area thick films of uniform thickness and of high green densities can be prepared by this technique. PZT thick films have applications in capacitors and as actuators.

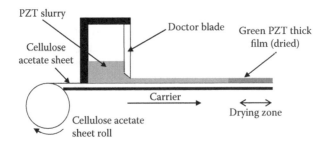

FIGURE 3.6
Tape casting technique for PZT thick films.

3.4.3 Fabrication of PZT Thin Films

Piezoelectric applications of PZT thin films include pressure sensors, micro-motors, and microvalves. They are also used for nonvolatile memory devices, capacitors, etc.

Thin films of PZT are prepared by the following techniques:

1. Vacuum sputter deposition technique
2. Spin coating technique
3. Chemical vapour deposition

3.4.3.1 RF Sputtering Deposition Technique

In the RF sputter deposition technique, plasma is generated in a low-pressure chamber using an inert gas, such as argon, by applying a high-frequency RF voltage across two electrodes. The material to be deposited (PZT) in the form of a plate is suitably positioned as the target in the chamber. The energetic ions of the plasma strike the target plate (PZT plate) and dislodge the molecules from the target. The high-frequency AC voltage is coupled capacitively through the ceramic target plate to the plasma so that conductive coating is not necessary on the target material. The dislodged molecules from the target travel through the plasma as vapour and get deposited on the substrate. Sputtering deposition is carried out at a temperature of not more than 150°C. The advantage of the sputtering technique over the thermal evaporation technique is that the stoichiometric ratio of the compound is maintained in the sputtering technique.

3.4.3.2 Spin Coating Technique

The spin coating technique is a relatively simple technique for growing thin films of PZT. The equipment consists of a variable speed spinning table. The substrate is fixed on the table by means of vacuum sucking. A small amount of PZT slurry is placed at the centre of the substrate using a nozzle. The substrate is made to spin at high speeds (500–5000 rpm), which makes the ceramic slurry spread uniformly on the substrate, forming a thin film. Subsequently, the film is baked to evaporate the solvent and cure the film. The thickness of the film is determined by the speed of rotation and the quantity of the liquid. Thin films of thicknesses in the range of 0.5–20 μm or thick films (5–100 μm) may be obtained by this method.

3.4.3.3 Chemical Vapour Deposition Technique

In the chemical vapour deposition technique, a chemical reaction is initiated in a vacuum chamber. The resulting chemical species get deposited on the substrate, which is heated to a high temperature (above 300°C). Normally,

CVD is carried out at atmospheric pressure. There are other variations of CVD, that is, low-pressure CVD (LPCVD) and plasma-enhanced CVD (PECVD), which are better suited for ceramic coatings. The steps involved in chemical vapour deposition techniques are

- Vapourization and transport of precursor molecules into the vacuum chamber (reactor)
- Diffusion of the molecules to the surface of the substrate
- Adsorption of the molecules to surface
- Decomposition of the molecules on the surface and formation of solid films

The CVD technique used for preparation of PZT films is normally referred to as the MOCVD technique. MO refers to the type of precursors used, which are compounds containing metal atoms linked to organic groups. Several precursors have been used for the preparation of PZT films, for example, tetraethyl lead (for lead), zirconium tetrabutoxide (for zirconium), and titanium iso propoxide (for titanium). The precursor vapours are fed into the reactor chamber, where the substrate (e.g., $Pt-TiO_2/SiO_2/Si$) is kept on a rotating disc and heated to a temperature of about 700°C. Films of different thicknesses in the range of 50–1400 nm have been grown by this technique.

3.5 Polymer Piezoelectric Materials

Polymer piezoelectric materials have special advantages over ceramic PZT. They are flexible, mechanically more stable, and can be obtained in the form of large-area thin films. Other advantages of polymer piezoelectric materials are they can be easily manufactured at much lower temperatures and can be formed more easily into different shapes. These properties make them attractive for piezoelectric applications, although the few known piezoelectric polymers have much lower piezoelectric coefficients compared to ceramic piezoelectrics. They have limited applications as actuators but are widely used as sensors.

The best polymer piezoelectric materials that are widely studied and used in applications are polyvinylidene fluoride (PVDF) and its copolymers. Applications include actuators for vibration and noise control, artificial muscles, and microfluidic components.

3.5.1 Polyvinylidene Fluoride (PVDF)

The molecular formula of PVDF is $[C_2H_2F_2]_n$. The structure of the molecule is shown in Figure 3.7a. The polymer exists in three different phases: α, β, and

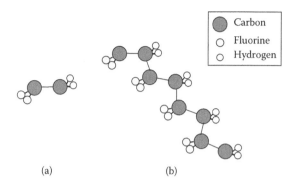

(a)   (b)

FIGURE 3.7
PVDF: (a) monomer, (b) polymer—phase.

γ phases. PVDF in β and γ phases are polar, while in α phase it is nonpolar. The structure of the β phase polymer is shown in Figure 3.7b. For piezoelectric applications, thin films of PVDF in the highly polar β phase are used.

PVDF thin films are formed either from the melt or from solution [10,11]. The phase of the film formed depends on the method used for solidifying the melt or the type of solvent used. When the melt (temperature >167°) is quenched to 80°C, α and β phases are formed. When α phase PVDF film is stretched to 300% strain, the film transforms to the highly polar β phase. The solvent cast method is used for the preparation of PVDF from solutions. The solvents used are dimethyl formamide or dimethyl acetamide. PVDF films are poled, keeping the film under stretched condition. A high electric field of the order of 20 kV/mm is applied for poling the films. The piezoelectric coefficient of PVDF film is in the range of 20–30 pC/N, which is ten times lower than that of PZT.

Copolymers of PVDF with trifluoroethylene and tetrafluoroethylene exhibit better piezoelectric properties than pure PVDF. The piezoelectric coefficient of the copolymer P(VDF-TrFE) is greater than 100 pC/N.

Because of their low stiffness constant, the polymers are not very useful as actuators, but they are useful as pressure sensors. For hydrophones (under water sonar detectors), the polymers are more suited than PZT because of their excellent impedance matching with water. Other common applications include pressure sensors in diesel injection lines and shock wave sensors.

3.6 Other Piezoelectric Materials

Piezoelectric materials described in the previous sections are the most commonly used materials for practical piezoelectric devices. There are many other piezoelectric materials, both natural and synthetic, which exhibit piezoelectric properties because of their noncentrosymmetric crystal structure. But only a few of them have been exploited for device applications. Examples of piezoelectric materials of significance are barium titanate and zinc oxide thin films. Other piezoelectric materials that are less important for device applications include tourmaline (natural mineral), lithium niobate, lead titanate, Rochelle salt, etc.

3.6.1 Barium Titanate

Barium titanate is a ferroelectric material with perovskite crystal structure. Although it exhibits reasonably good piezoelectric properties, it is not as widely used as PZT for actuator and transducer applications. Its piezoelectric applications are limited to microphones and buzzers. Nanocrystalline barium titanate has a high dielectric constant and is useful as a dielectric material in ceramic capacitors. The ceramic exhibits a high positive temperature coefficient of resistance, which makes it useful as a temperature measuring and controlling device (thermistor) in heating systems.

3.6.2 Zinc Oxide Thin Films

The crystal structure of ZnO is hexagonal wurtzite with 6 mm symmetry. Due to this symmetry, a single crystal of ZnO has a unique direction (C-axis), which is the polar direction. ZnO thin films with high C-axis orientation exhibit good piezoelectric properties.

ZnO crystals in the form of thin films or other nanostructural forms such as nanowires and nanotubes have been studied extensively for various piezoelectric applications. Some of the techniques used for the preparation of ZnO nanostructures are RF sputtering, sol–gel spin coating, chemical vapour deposition, and pulsed laser deposition [12–14]. ZnO nanocrystals grown in the form of thin fibres have been investigated for applications as nanogenerators [15,16]. Synthesis of ZnO crystals in the form nanobelts, nanorings, and nanosprings that could be used for nanoactuators and sensors is reported by Xiang Yang Kong and Zhong Lin Wang [17]. The fabrication of piezoelectric ZnO thin films for application as ultrasonic transducer arrays operating at 100 MHz is reported by Y. Ito et al. [18]. Fabrication of ZnO microcantilever using micromachining techniques for application as atomic force microscope probes is described by R. L. Johnson [19].

3.7 Composite Piezoelectric Materials

The majority of piezoelectric devices use the ceramic PZT. Ceramic materials are by nature mechanically weak because of their brittleness. Applications of

PZT in transducers are limited to devices in which the stress on the materials is below a certain limit. In certain applications such as accelerometers and vibration sensors in aerospace and transducers in naval applications, more robust and mechanically more stable materials are required. Composites made of ceramic PZT combined with a polymer material have been found to possess several improved characteristics compared to single-phase ceramic PZT. Although piezoelectric sensitivities of composites are much less than those of single-phase PZT, composites have several advantages in specific applications. Besides being mechanically more stable, they offer more versatility in the design and show better performance in applications such as vibration sensors, shock accelerometers, underwater transducers (hydrophones), etc.

Composite piezoelectric materials, more often referred to as "piezocomposites," are composed of more than one material component, mostly not more than two. Two-phase piezocomposites are composed of a piezoelectric ceramic material (mostly, PZT) and a polymer material; the polymer material may be piezoelectric (e.g., PVDF) or may be any other nonpiezoelectric polymer (e.g., epoxy resin). The combination of ceramic and polymer provides a mechanically more stable material because of the high compliance of the polymer phase. Two-phase PZT–polymer composites have been widely studied and reported in the literature [20–28] and even commercially exploited by piezoelectric transducer manufacturers.

Piezocomposites are classified based on the manner in which the two phases are combined in the material. The notation used for the type of combination is n_1-n_2, where n_1 and n_2 are integers taking values 1, 2, or 3. The numbers specify the number of directions in which phase 1 (ceramic) and phase 2 (polymer) are continuously connected in 3 dimensions. The practical combinations are 0-3, 1-3, and 3-3, of which 0-3 and 1-3 are studied and reported most.

The properties of the composites are governed by following factors:

- Properties of the individual phases
- Volume fraction of the phases
- Type of connectivity pattern
- Technique used for the fabrication of the composite

3.7.1 0-3 PZT–Polymer Composites

In this type, fine powders of ceramic PZT are dispersed uniformly in a continuous polymer matrix. PZT is disconnected in all the three directions ($n_1 = 0$), but the polymer is continuous and so connected in all the three directions ($n_2 = 3$). This combination is shown in Figure 3.8.

One of the methods used for the fabrication of 0-3 PZT composites is the solvent–cast technique. In this technique, a measured quantity of the

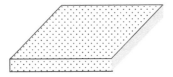

FIGURE 3.8

PZT–polymer composite sheet with 0-3 connectivity. Fine powders of PZT are uniformly dispersed in a continuous polymer matrix.

polymer is dissolved in a suitable solvent. A proportionate quantity of PZT powder is mixed thoroughly with the polymer solution, and the mixture is heated until it becomes viscous. The solution is then cast on glass plates, and the solvent is allowed to evaporate. Thin sheets of 0-3 connectivity composites are formed, which can be peeled out of the glass plates. The drawback of this method is that the thickness of the sheets cannot be controlled and the density of the sample will be less than the expected theoretical density.

Another technique successfully used for the preparation of 0-3 composites is the hot-press technique. In this technique, the polymer solution with the PZT powder is heated to get a thick viscous mixture. The mixture is poured into distilled water in a beaker at room temperature. It is then filtered and dried. The dried mixture of PZT–polymer is then heated until the solvent completely evaporates and is further heated to melt the polymer. The mixture is then transferred to a die and pressed under controlled pressure and temperature. In this technique, 0-3 composites of desired shapes such as thick square plates or circular discs or cylinders can be obtained.

The injection moulding technique has also been used to fabricate 0-3 connectivity PZT–polymer composites. In this technique, fine powders of PZT are mixed well with the polymer in powder form in a rotary mixer, and the mixture is injection-moulded at high temperature. The resulting composite in the required shape is then cured.

3.7.2 1-3 PZT–Polymer Composites

In this type of composite, closely spaced thin rods or fibres of PZT are reinforced in a continuous polymer matrix with all the fibres parallel to each other and across the thickness of the matrix ($n_1 = 1$). The polymer is continuous and so connected in all the three directions ($n_2 = 3$). This combination is schematically shown in Figure 3.9A.

A commonly used technique for the fabrication of 1-3 composite is the injection moulding technique. In this technique, PZT powder is mixed with an organic binder and is heated. The hot mixture is injected into a mould that is shaped suitably for 1-3 configuration. The product is removed from the mould and heated to remove the binder. The ceramic is then sintered. The vacant spaces in the ceramic preform are filled with the polymer (epoxy)

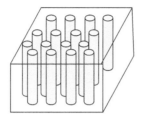

FIGURE 3.9
1-3 PZT–polymer composites. A. 3-D schematic of 1-3 PZT–polymer composite.

PZT

PZT

Polymer

(a)   (b)

FIGURE 3.9
B. 1-3 composite by injection moulding technique: (a) cross section of PZT ceramic preform, (b) cross section of 1-3 PZT–polymer composite.

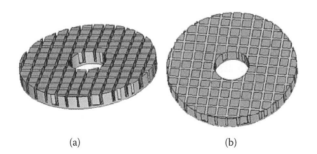

(a)   (b)

FIGURE 3.9
C. 1-3 composite by dice-and-fill technique: (a) PZT disc with grooves, (b) grooves in PZT disc filled with polymer. (Courtesy: Marco Aurélio B. Andrade et al., 2009, *J. Braz. Soc. Mech. Sci. Eng.* Copyright © 2009 by ABCM October–December, 2009, Vol. XXXI, No. 4/313.)

by the usual encapsulation technique. After encapsulation, the ceramic base is removed by grinding. The PZT ceramic preform and the 1-3 composite are shown in Figure 3.9B.

Another technique used for fabrication of 1-3 composites is the dice-and fill-technique. In this technique, grooves are made as required in a PZT disc or plate using a diamond-cutting machine. The grooves in the ceramic are then filled with the polymer, and the system is degassed in a vacuum chamber. The composite disc is then cured at room temperature for several hours. The PZT discs with grooves and after filling with the polymer are shown in Figure 3.9C.

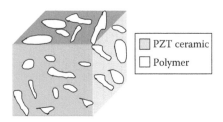

FIGURE 3.10
3-3 Ceramic–polymer composite.

3.7.3 3-3 PZT–Polymer Composite

In this type of composite, the two phases are continuously self-connected in all the three directions. A highly porous PZT ceramic material can be thought of as a 3-3 composite of the two phases, PZT and air. Such porous PZT ceramic samples can be prepared by impregnation of polyethylene foam or wax replamine of coral reef with the ceramic slip and heating it to remove the foam or the wax. Later, the sample is sintered at a high temperature. The porous spaces in the ceramic sample are then impregnated with a polymer material (Figure 3.10).

Composites of 3-3 connectivity have advantages over single-phase ceramic PZT, especially in applications such as hydrophones (sonar receivers) and ultrasonic imaging devices because of their low density, low stiffness constant, and high hydrostatic sensitivity. The composites have better acoustic impedance matching with water and human tissues.

References

1. K. P. Rema et al., 2009, Influence of low lanthanum doping on the electrical characteristics of PZT 53/47, *Journal of Physics D: Applied Physics*, Vol. 42, No. 7. 075420-6 pp.
2. V. Singh et al., 2006, Effect of lanthanum substitution on ferroelectric properties of niobium doped PZT ceramics, Vol. 60, No. 24, 2964–2968.
3. B. Sahoo et al., 2006, Development of PZT powders by wet chemical method and fabrication of multilayered stacks/actuators, *Materials Science and Engineering B*, Vol. 126, 80–85.
4. P. S. Gaware et al., 2003, Manufacturing technology of lead zirconate titanate cylindrical elements for passive transducer arrays, *Defence Science Journal*, Vol. 53, No. 3, 275–279.
5. B.-H. Chen, 2010, Fabrication of PZT by sol-gel method, Symposium on Piezoelectricity, Acoustic Waves and Device Applications, Xiamen.

6. G. Mu et al., 2007, Synthesis of PZT nanocrystalline powder by a modified sol-gel process using water as primary solvent source, *Journal of Materials Processing Technology*, Vol. 182, No. 13, 382–386.

7. A. Sen et al., 2005, Technological challenges of making PZT based piezoelectric wafers, *Proceedings of International Conference on Smart Materials Structures and Systems*, July 28–30, Bangalore, India.

8. A. Zarycka et al., 2003, Application of the sol-gel method to deposition of thin films, *Materials Science*, Vol. 21, No. 4, 439–443.

9. C.-C. Chang, 2000, The fabrication and characterization of PZT thin film acoustic devices for application in underwater robotic systems, *Proceedings of the National Science Council*, ROC(A), Vol. 24, No. 4, 287–292.

10. S. Bauer, 2006, Piezoelectric polymers, *Materials Research Society Symposium Proceedings*, Vol. 889, Materials Research Society.

11. D. M. Esterly, 2002, Manufacturing of Poly(vinylidene fluoride) and Evaluation of its Mechanical Properties, M. S. thesis, Virginia Polytechnic Institute and State University.

12. K.-M. Zhang et al., 2007, Piezoelectricity of ZnO films prepared by sol-gel method, *Chinese Journal of Chemical Physics*, Vol. 20, No. 6, 721–726.

13. S. Ilican et al., 2008, Preparation and characterization of ZnO thin films deposited by sol-gel spin coating method, *Journal of Optoelectronics and Advanced Materials*, Vol. 10, No. 10, 2578–2583.

14. Lamia Znaldi, 2010, Sol-gel-deposited ZnO thin films: A review, *Materials Science and Engineering B*, Vol. 174, No. 1-3, 18–30.

15. T.-H. Fang and S.-H. Kang, 2010, Physical properties of ZnO: Al nanorods for piezoelectric nanogenerator application, *Current Nanoscience*, Vol. 6, No. 05, 505–511.

16. Z. L. Wang, 2009, Energy harvesting using piezoelectric nanowires, *Advanced Materials*, 21, 1311–1315.

17. X. Y. Kong and Z. L. Wang, 2003, Spontaneous polarization-induced nanohelixes, nanosprings, and nanorings of piezoelectric nanobelts, *Nano Letters*, 3 (12), 1625–1631.

18. Y. Ito et al., 1995, A 100 MHz ultrasonic transducer array using ZnO thin films, IEEE Transactions on Ultrasonics, Ferroelectrics and Frequency Control. Vol. 42, No. 2, 316–324.

19. R. L. Johnson, Characterization of Piezoelectric ZnO Thin Films and Fabrication of Piezoelectric Micro-Cantilevers, M. S. thesis, Iowa State University.

20. T. R. Gururaja, A. Safari, R. E. Newnham, and L. E. Cross, 1987, *Piezoelectric Ceramic-Polymer Composites for Transducer Applications, Electronic Ceramics*, Ed. L. M. Levinson, Marcel Dekker, New York, pp. 92–128.

21. R. E. Newnham, D. P. Skinner, and L. E. Cross, 1978, Connectivity and piezoelectric-pyroelectric composites, *Materials Research Bulletin*, Vol. 13, No. 5, 525–536.

22. R. E. Newnham, A. Safari, J. Giniewicz, and B. H. Fox, 1984, Composite piezoelectric sensors, *Ferroelectrics*, Vol. 60, 15–21.

23. T. R. Shrout, W. A. Schulze, and J. V. Biggers, 1979, Simplified fabrication of PZT/polymer composites, *Materials Research Bulletin*, Vol. 14, No. 12, 1553–1559.

24. S. Y. Lynn, R. E. Newnham, K. A. Klicker, K. Rittenmyer, A. Safari, and W. A. Schulze, 1981, Ferroelectric composites for hydrophones, *Ferroelectrics*, Vol. 38, 955–958.

25. J. F. Tressler et al., 1999, Functional composites for sensors, actuators and trans-ducers, *Composites: Part A*, Vol. 30, No. 4, 477–482.
26. K. Rittenmyer, T. Shrout, W. A. Schulze, and R. E. Newnham, 1982, Piezoelectric *3-3* composites, *Ferroelectrics*, Vol. 41, 189–195.
27. Y.-C. Chen and S. Wu, 2001, Piezoelectric composites with 3-3 connectivity by injecting polymer for hydrostatic sensors, *Journal of Electroceramics*, Vol. 8, No. 3, 209–214.
28. V. Janas and A. Safari, 1995, Overview of fine-scale piezoelectric ceramic/polymer composite processing, *Journal of the American Ceramic Society*, Vol. 78, No. 11, 2945–2955; American Ceramic Soc., Westerville, USA.

4

Engineering Applications
of Piezoelectric Materials

4.1 Introduction

Piezoelectric materials are inherent transducers that convert mechanical energy to electrical energy (direct piezoelectric effect) and electrical energy to mechanical energy (indirect piezoelectric effect). Based on these effects, the materials have multitude of applications in the field of engineering and medicine. The unique characteristic of the material of possessing energy conversion capability in two directions, that is, from mechanical to electrical and from electrical to mechanical, makes piezoelectric materials highly suited for design of smart systems.

Engineering applications of piezoelectric materials are summarized in Table 4.1, and each of the applications is discussed in the following sections of this chapter.

4.2 Gas Lighter

A gas lighter is a common piezoelectric household device which makes use of the direct piezoelectric effect to generate electric sparks.

In a gas lighter, a high voltage pulse is required to be generated across a narrow electrode gap. A piezoelectric gas lighter consists of a PZT cylinder which is subjected to a stress pulse using a spring mechanism. When a button is pressed, a stress pulse is applied on the piezoelectric cylinder. The stress causes a high voltage to be generated which is made to appear across a small air gap between two closely spaced electrodes. The arrangement is shown in Figure 4.1. The voltage developed is high enough to cause breakdown of the air gap between the two electrodes, resulting in a spark.

TABLE 4.1

Engineering Applications of Piezoelectric Materials

Piezoelectric Effect Used	Energy Conversion	Applications
Direct effect	Input: Mechanical Output: Electrical	Gas lighter Pressure sensor Accelerometer Gyroscope (rotation sensor) Piezoelectric microphone Ultrasonic detector Hydrophone (SONAR) Tactile sensor Energy harvesting
Indirect effect	Input: Electrical Output: Mechanical	*Low-frequency applications* 　Electronic buzzer 　Tweeters (high-frequency speakers) 　Actuators *High-frequency applications* 　Piezoelectric motor 　Piezoelectric pump 　Ultrasonic drill 　Ultrasonic cleaner 　Ultrasonic generator 　Projector (SONAR)
Both direct and indirect effects	Input: Electrical/ Mechanical Output: Mechanical/ Electrical	Quartz crystal oscillator Quartz crystal balance Quartz crystal AFM probe Piezoelectric transformer Ultrasonic nondestructive testing Noise and vibration control Structural health monitoring Smart devices and robots

A stress pulse of amplitude X_3 generates an electric field E_3 (given in Chapter 2, Equation 2.4):

$$E_3 = g_{33}X_3 \tag{4.1}$$

The electric field is generated across the length l of the cylinder. The open circuit voltage generated will be

$$V = lE_3 = l\,g_{33}X_3 = l\,g_{33}\frac{F}{A} \tag{4.2}$$

where F is the force pulse amplitude and A is the area of cross section of the cylinder.

Taking the g_{33} value of piezoelectric material (PZT) as 22×10^{-3} Vm/N, the voltage generated across a cylinder of length 1 cm and radius 5 mm for an

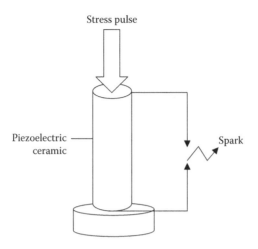

FIGURE 4.1
Piezoelectric gas lighter.

applied force of 4.5×10^3 N would be about 12.5 KV, which is high enough to cause a spark.

4.3 Pressure Sensor

Piezoelectric pressure sensors make use of the direct piezoelectric effect. The pressure to be measured is made to fall on a piezoelectric membrane which generates electric voltage proportional to the input pressure. Piezoelectric pressure sensors have been designed using quartz, PZT, and ZnO thin films. The pressure sensors are very sensitive and can be used to measure absolute pressures.

Quartz piezoelectric pressure sensors are quite common, and they are used in industries for measuring cylinder pressures in internal combustion engines and for measuring pressure changes in pneumatic and hydraulic systems. The sensing element in the pressure sensor is a stack of thin synthetic quartz crystals. The pressure to be measured acts on a diaphragm which generates compressive force on the quartz crystals stack. A proportionate analog voltage is generated by the quartz crystals. The voltage signal is measured using suitable electronic circuits. Some of the commercial quartz crystal pressure transducers are Intertechnology Inc, Dytran Instruments, and PCB Piezotronics. A quartz crystal pressure sensor marketed by Intertechnology Inc. is shown in Figure 4.2.

PZT thin films have been used to design MEMS pressure sensors. A typical MEMS PZT pressure sensor consists of a silicon substrate (of about 500 μm

FIGURE 4.2
A quartz crystal piezoelectric pressure sensor. Courtesy: Intertechnology Inc.—Testing and Measurement Solutions, Ontario, Canada.

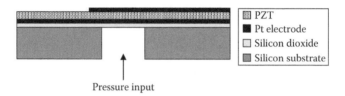

Pressure input

FIGURE 4.3
Schematic of a MEMS PZT pressure sensor.

thickness) with a silicon oxide coating of about 100–200 nm thickness. On the silicon oxide layer, a sandwich of Pt-PZT-Pt structure is grown. The platinum layers act as electrodes, and the PZT is the active material that senses the pressure. The PZT film thickness is usually in the range of 0.5–5 μm. The schematic of the structure of a MEMS piezoelectric pressure sensor is shown in Figure 4.3. Pressure sensors which can measure up to 40,000 psi have been designed for military applications using PZT thin films [1].

Piezoelectric pressure sensors using ZnO thin films have been designed and reported in the literature. ZnO thin film formed on a thin membrane of polyamide bonded to a silicon wafer has been designed as a differential pressure liquid flow sensor [2].

4.4 Accelerometer

Accelerometers are used for measurement of vibrations in many applications which include impact acceleration levels experienced by vehicles during crash, shock experienced by space vehicles and cargo during stage separation, testing of shock resistance of packaged products, vibrations in mining activities, seismic vibrations during earthquakes, etc.

4.4.1 Principle of Piezoelectric Accelerometer

A piezoelectric accelerometer basically consists of a piezoelectric disc clamped between a base plate and a seismic mass *M* as shown in Figure 4.4.

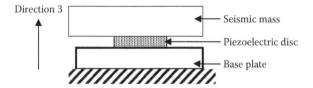

FIGURE 4.4
Schematic diagram of basic parts of a typical compressive type piezoelectric accelerometer.

When the system is subjected to acceleration, the seismic mass exerts a force F on the piezoelectric disc given by

$$F = Ma \tag{4.3}$$

where a is the acceleration experienced by the disc. The mechanical stress in direction 3 on the piezoelectric disc is given by

$$X_3 = \frac{F}{A} = \frac{Ma}{A} \tag{4.4}$$

where A is the area of the disc. The mechanical stress causes an electric field E to be generated across the thickness of the disc given by (Equation 2.4)

$$E_3 = g_{33} X_3 \tag{4.5}$$

The open circuit voltage across the piezoelectric disc of thickness t will be

$$V = E_3 t = g_{33} X_3 t \tag{4.6}$$

Substituting for X_3 from Equation 4.4,

$$V = g_{33} \frac{t}{A} Ma \tag{4.7}$$

The output voltage is proportional to the acceleration. The proportionality constant is determined by the seismic mass M, the piezoelectric coefficient g_{33}, and the dimensions of the piezoelectric disc.

If the accelerometer is subjected to vibrations, it can be treated as a mass–spring system and can be represented by an electrical equivalent circuit shown in Figure 4.5.

The inductance L_m represents seismic mass, the capacitance C_m represents the stiffness constant, and C_o is the electrostatic capacitance of the PZT disc. For frequencies far below the resonance frequency of the system, the inductance L_m representing the seismic mass may be neglected. The accelerometer

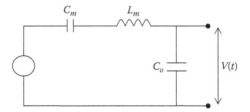

FIGURE 4.5
Electrical equivalent circuit of a piezoelectric accelerometer.

is then represented by the sum of the electrostatic capacitance of the piezo-electric disc C_o and the mechanical capacitance C_m:

$$C_a = C_o + C_m$$

The accelerometer output is connected to a preamplifier. The open circuit voltage will be maintained if the input capacitance C_E and the resistance R_E of the amplifier satisfy the conditions

$$C_E \ll C_a$$

and

$$R_E \gg \frac{1}{\omega C_a}$$

The output of the preamplifier is a measure of the acceleration. The sensitivity of the accelerometer is expressed as the output voltage per unit of input acceleration. The unit of acceleration is taken as g, the acceleration due to gravity. Sensitivity is expressed as millivolts per g of acceleration. The output voltage may be RMS voltage or peak voltage.

4.4.2 Operating Frequency Range

Operating frequency range is the frequency range over which the sensitivity of the transducer remains almost a constant. This range is determined by the mechanical and electrical characteristics of the transducer. A typical sensitivity versus frequency characteristic of a piezoelectric accelerometer is shown in Figure 4.6.

Low-frequency response does not depend on the accelerometer system but depends on the input impedance of the preamplifier. The low-frequency limit is a function of the RC time constant formed by the accelerometer cable and the input capacitance and resistance of the amplifier. The high-frequency limit depends on the resonance frequency of the accelerometer. The sensitivity is almost a constant up to about 1/5 of the resonance frequency. To

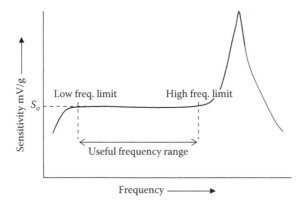

FIGURE 4.6
Typical frequency characteristic of a piezoelectric accelerometer.

increase the useful range of the accelerometer, the resonance frequency must be shifted to higher values. This can be achieved by reducing the seismic mass; but reducing the seismic mass lowers the sensitivity of the accelerometer. The seismic mass is chosen such that there is a compromise between the frequency range and sensitivity. In shock accelerometers, normally the high-frequency range is required, and so a lower seismic mass is used. Frequency responses for two commercial piezoelectric accelerometers with different seismic masses are compared in Figure 4.7 [3]. The one with the lower seismic mass has lower sensitivity but a higher useful frequency range.

Environmental effects that may affect the accelerometer sensitivity are temperature and humidity. Temperature effects may be compensated by using suitable temperature control circuits, and the effect of humidity may be reduced by hermetically sealing the device.

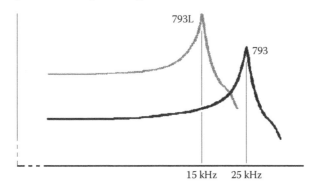

FIGURE 4.7
Frequency response of two piezoelectric accelerometers of different seismic masses. The 793 accelerometer has a lower seismic mass than that of the 793L. (Courtesy: Wilcoxon Research, MEGGITT, Industrial accelerometer design. Taken from the website: http://www.wilcoxon. com/knowdesk/Industrial%20piezoelectric%20accelerometer%20design.pdf.)

4.4.3 Types of Piezoelectric Accelerometers

Accelerometers may be (1) compression type, (2) shear type, or (3) bending type. Designs of the three types are shown in Figure 4.8.

A compression-type accelerometer consists of a piezoelectric disc with a central hole fitted to a shaft. The disc is attached to a rigid base, and a seismic mass is placed on top of the disc. The disc and the seismic mass are firmly bonded to each other. The system is mounted in a rigid frame with a spring as shown in Figure 4.8A [4]. The external force due to vibration causes compressive forces to act on the piezoelectric disc along its thickness,

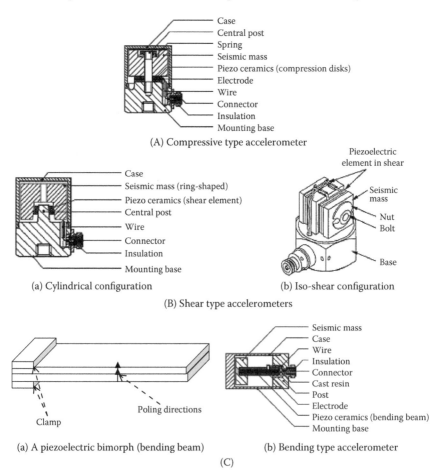

FIGURE 4.8
Types of accelerometers. (A) Compressive-type accelerometer, (B) (a) shear-type accelerometer—cylindrical configuration, (C) (b) Bending-type accelerometer. (Courtesy: Piezoelectric Accelerometers, Theory and Applications, Metra Mess-und Frequenztechnik, 2001.) (B) (b) Iso-shear accelerometer. (Courtesy: Harris' *Shock and Vibration Handbook*, 5th edition, Eds: Cyril M. Harris, Allan G. Piersol, McGraw-Hill, New York, 2002. Chapter 12 by Anthony S. Chu.)

the force being proportional to the seismic mass. Due to the longitudinal piezoelectric effect, charges are generated on the electroded faces of the disc. The piezoelectric coefficient involved is d_{33}. The charge is measured using a charge-to-voltage converter and a preamplifier. Usually, PZT is used for the piezoelectric material.

In shear-type accelerometers, the piezoelectric elements experience a shear force due to the input acceleration. The piezoelectric coefficient involved is d_{15}. One type of shear accelerometer consists of a piezoelectric cylindrical ring fitted around a cylindrical shaft with a rigid base. A ring-shaped seismic mass is mounted on top and firmly bonded to the piezoelectric ring (Figure 4.8B(a)) [4]. When the system is subjected to an external force, a shear force acts on the face of the piezoelectric cylinder, and the charges generated are measured.

Another type of shear accelerometer called the iso-shear accelerometer consists of two square flat piezoelectric plates fitted on either side of a flat metal plate using a nut and bolt as shown in Figure 4.8B(b) [5]. The piezoelectric elements are prestressed. The assembly is mounted on a strong base plate as shown. When the accelerometer is subjected to external vibrations, the two piezoelectric plates experience shear strain, and charges are generated on the faces, which are measured using suitable electronic circuits. The sensitivities of the accelerometers are in the range of 10–500 picocoulmbs/g. The useful frequency range is about 3–5000 Hz.

In the bending-type accelerometer, a piezoelectric bimorph which bends when subjected to external acceleration is used. A piezoelectric bimorph consists of two thin piezoelectric strips bonded to each other as shown in Figure 4.8C(a). The system is clamped at one end and free at the other end. The two strips are poled in the same direction as shown. When the system is subjected to bending due to acceleration, the top strip expands, and the bottom strips contracts. The piezoelectric coefficient involved is d_{13}. The two strips acquire positive charges on the outer faces and negative charges on the common face. The bending-type accelerometer which uses the bimorph as the piezoelectric element is shown in Figure 4.8C(b) [4]. These types of bending accelerometers are quite sensitive, but the disadvantage is that they are more fragile, and the useful range of frequencies is less than those of the other types.

4.5 Piezoelectric Gyroscope–Angular Rate Sensors

Gyroscopes are devices used for measuring the angular rate of rotating objects. They have several engineering applications: in automobiles they are used for stability control, navigation assistance, and rollover detection; in marine engineering, they are used for stabilization and navigation of ships and in military applications for missile stabilization and guidance. There are mainly two types of gyroscopes: (1) spinning mass gyroscopes and (2) vibratory gyroscopes.

Vibratory gyroscopes have the advantage that they do not need motors or bearings. They are more rugged and have a better lifetime. Vibratory gyroscopes are classified based on the actuation mechanisms used. The different actuation mechanisms used are electrostatic, electromagnetic, and piezoelectric. In piezoelectric vibratory gyroscopes, piezoelectric material is used for both actuation and sensing. This section deals with piezoelectric vibratory gyroscopes, which are widely used because of their simplicity in design and high sensitivity.

The principle used in all types of gyroscopes is the Coriolis effect, which arises in a rotating frame of reference. The Coriolis effect may be stated as follows: "When a moving object is subjected to rotation about an axis perpendicular to the direction of motion, the object experiences an acceleration in a direction mutually perpendicular to the original direction of motion and the axis of rotation." The equation that describes the Coriolis effect is

$$\vec{a} = -2\vec{\Omega} \times \vec{\upsilon} \qquad (4.8)$$

where $\vec{\upsilon}$ is the initial velocity of the object, $\vec{\Omega}$ is the angular velocity of rotation, and $\vec{a}$ is the acceleration acquired.

In a vibratory gyroscope, an object is made to vibrate in a specific direction. If the object experiences a rotation in a perpendicular direction, it acquires vibrational motion in a third mutually perpendicular direction and causes a change in the vibration pattern of the object. This change is a measure of the applied rotation rate.

Piezoelectric vibratory gyroscopes use the piezoelectric effect for excitation and detection of vibrations in resonators. A piezoelectric actuator is used for excitation of vibration, and a piezoelectric sensor is used for detection. Vibrations that arise due to excitation are called the primary mode, and vibrations that arise due to the Coriolis effect when there is a rotation are called the secondary mode. The amplitude of the secondary vibration is proportional to the rate of rotation. The piezoelectric gyroscope is placed in a suitable location on the object whose rotation rate is to be measured. When the object experiences a rotation, the rotation rate is measured by detecting the vibration of the secondary mode. The principle of a vibratory gyroscope is illustrated in Figure 4.9.

The primary mode of vibration is along the X-direction. The rotation of the object is anticlockwise with the axis of rotation in the Z-direction. The secondary mode of vibration acquired due to the Coriolis effect is in the Y-direction as shown. In a piezoelectric gyroscope, the primary mode is induced using a piezoelectric actuator, and the secondary mode is detected using a piezoelectric sensor.

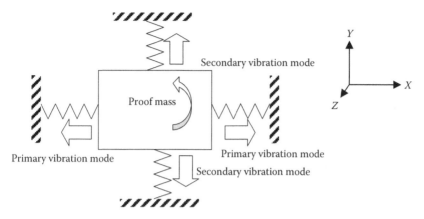

FIGURE 4.9
Schematic diagram of a vibratory gyroscope. The primary vibration mode is along the X-direction. The rotation of the proof mass is anticlockwise about the Z-direction. The secondary vibration mode is along the Y-direction.

Types of piezoelectric gyroscopes are

- Piezoelectric beam gyroscope
- Piezoelectric tuning fork gyroscope
- Piezoelectric disc, cylinder, or ring gyroscope

4.5.1 Piezoelectric Beam Gyroscope

A piezoelectric beam gyroscope consists of a steel beam of cross section (1) triangular, (2) rectangular, or (3) circular. Piezoelectric elements are attached to the faces of the beams. The beam is subjected to rotation about the axis of the beam. The three types of beams are shown in Figure 4.10.

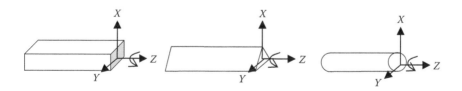

FIGURE 4.10
Piezoelectric beam gyroscopes: (a) beam of rectangular cross section, (b) beam of triangular cross section, (c) beam of circular cross section. Piezoelectric elements are bonded to the faces of the beams at one end (not shown in the figures). The beams rotate about the Z-axis. The vibrations are excited and detected in the X- and Y-directions.

The piezoelectric element on one of the faces of the beam, called the drive plane, is used to excite primary vibration in the beam. The beam is mounted such that the rotation is about the axis of the beam. The piezoelectric element on the face perpendicular to the drive plane senses the secondary vibration when the beam is subjected to rotation. Another piezoelectric element on the face parallel to the drive face is used as a feedback in the electronic driving circuit to keep the vibration amplitude constant at the mechanical resonance frequency of the beam.

4.5.2 Piezoelectric Tuning Fork Gyroscope

A piezoelectric tuning fork gyroscope is either made of a single crystal of quartz or made of any other material, and piezoelectric elements are bonded to the tines of the fork for actuation and detection. The tuning fork is excited by applying an external voltage to the electrodes on the tines. This is the primary vibration in the Y-direction (Figure 4.11). When the stem of the tuning fork rotates about the Z-axis, a secondary vibration of the tines is generated in the X-direction as shown. The amplitude of this vibration is measured as the rate of rotation.

A micromachined double-ended quartz tuning fork is schematically shown in Figure 4.12 [6]. The actuator tuning fork is excited by an oscillator circuit at the resonance frequency of the system. The tines vibrate toward and away from one another in the plane of the system.

When the system rotates about its axis, the Coriolis effect generates an oscillating force in a direction perpendicular to the rotation axis and the vibration direction. This makes the sensor tines vibrate in and out of the plane as shown, producing a signal which is measured as the rotation rate.

4.5.3 Piezoelectric Cylindrical Gyroscope

There are two types of piezoelectric cylindrical vibratory gyroscopes. One type uses a thin-walled steel cylindrical shell to which piezoelectric

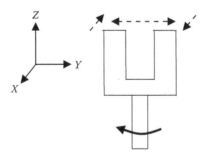

FIGURE 4.11
Principle of a tuning fork gyroscope. The tuning fork is excited to vibrate in the Y-direction. The rotation of the tuning fork along the Z-direction generates secondary vibration of the tines in the X-direction as shown.

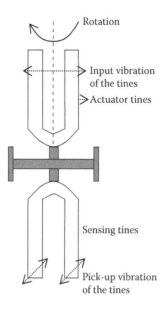

Rotation

Input vibration
of the tines

Actuator tines

Sensing tines

Pick-up vibration
of the tines

FIGURE 4.12
Schematic diagram of a double-ended quartz tuning fork gyroscope. The actuator tines are excited to vibrate in the plane towards and away from each other. When there is a rotation about the axis of the tuning fork system, the sensor tines vibrate in and out of the plane as shown.

elements are symmetrically attached for actuation and sensing [7]. In the second type, the cylindrical shell is made of a piezoelectric material, and suitable electrodes are provided on the walls of the cylinder for excitation and sensing [8–10].

The two types of gyroscopes are shown schematically in Figure 4.13.

The two modes of vibration of the cylinder used in the vibratory gyroscope are shown in Figure 4.14. These two modes are called the primary and the secondary modes. The primary mode has antinodes at 0°, 90°, 180°, and 270°. The secondary mode has the same form as the primary mode, but is rotated by 45° with respect to the primary mode.

The cylinder is clamped at one end, and the other end is free to vibrate. The cylinder is placed on the rotating object whose rotation is to be measured, such that the rotation axis is along the axis of the cylinder.

4.5.3.1 Operation of the Steel Cylinder Gyroscope

The faces of the piezoelectric elements touching the cylinder are grounded. The opposite pairs of elements are electrically connected. The arrangement is shown schematically in Figure 4.15. The piezoelectric elements at positions 1 and 3 are used for actuation and the elements at 2 and 4 are used for sensing the rotation.

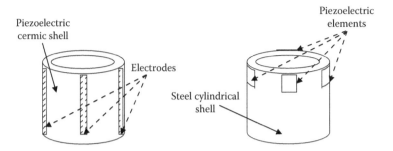

FIGURE 4.13
Schematic diagram of piezoelectric cylindrical shell gyroscopes: (a) Cylindrical shell made of piezoelectric ceramic material electroded in required parts, (b) cylindrical shell of steel to which four piezoelectric elements are attached.

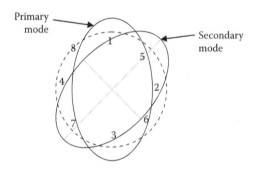

FIGURE 4.14
The two modes of vibration of a cylindrical shell: (i) Primary mode has antinodes at points 0° (at 1), 90° (at 2), 180° (at 3), and 270° (at 4), (ii) secondary mode has antinodes at points shifted by 45° with respect to the primary antinodes (at points 5, 6, 7, and 8).

A sinusoidal voltage of frequency equal to the resonance frequency of the cylinder is applied to the actuator elements at 1 and 3. This makes the cylinder vibrate at its resonance frequency. When a rotation rate is applied to the cylinder about its axis, the Coriolis force couples the energies of the primary mode and the secondary mode. This results in a change of voltage appearing at the piezoelectric sensors at positions 2 and 4. The change in voltage at the sensors is proportional to the rate of rotation. The sensor voltage can be calibrated to measure the rotation rate of the cylinder. The two sensor elements may be shorted, and the output may be measured or the differential output of the two sensors may be measured.

Finite element analysis of the cylindrical gyroscope described above is given in Chapter 6 (Section 6.5.3).

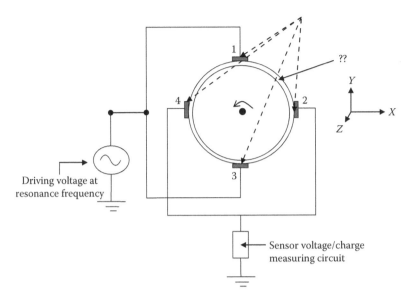

FIGURE 4.15
Piezoelectric cylindrical shell gyroscope with four piezoelectric elements. Primary vibration is actuated with the two piezoelectric elements at 1 and 3. When the cylinder rotates about the Z-axis, the change in vibration mode is detected at the two sensor piezoelectric elements at 2 and 4.

4.6 Piezoelectric Microphone

The piezoelectric microphone is based on the direct piezoelectric effect, that is, conversion of sound vibrations to electrical signals. The microphones are made of thin plates of quartz crystals or PZT thin films. The dimensions of the plates or films are selected to have sensitive output signals in the audible sound frequency range. In a piezoelectric MEMS microphone, PZT thin films are formed on silicon substrate, and micromachining techniques are used to form the membrane. A preamplifier is built into the system. A schematic of a MEMS piezoelectric microphone is shown in Figure 4.16.

4.7 Piezoelectric Ultrasonic Transducers for Sound Navigation and Ranging (SONAR)

Piezoelectric discs are capable of vibrations at very high frequencies in the ultrasonic range. They can be used to generate and detect ultrasound, which is in the frequency range of 20 kHz and up to several MHz.

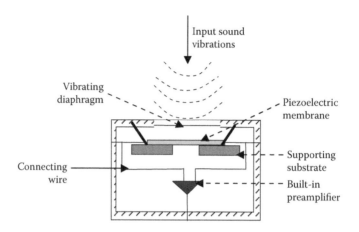

FIGURE 4.16

Piezoelectric microphone. A thin piezoelectric plate (quartz) or a PZT membrane vibrates due to input sound vibrations. The output signal from the piezoelectric material is amplified with a built-in preamplifier circuit.

The SONAR technique involves generation and transmission of ultrasound in water and receiving and sensing the ultrasound reflected from submerged objects. Since piezoelectric materials can be used for both generation and detection of sound, the same device can be used as both ultrasonic generator and detector. In underwater acoustics, sound generators are called *projectors*, and sound detectors are called *hydrophones*.

Sound ranging and detection under water has several applications which include

- Measurement of noise generated by vessels at various speeds. The noise levels are required to be maintained within the stipulated limits. The testing is done for new vessels or whenever new equipments are installed in the vessels.
- Underwater communication
- Detection and ranging of enemy vessels under water (submarines).
- Oceanographic survey: Study of the sub-bottom profile of the sea by offshore oil industries
- Marine mammal research

Hydrophones are ultrasonic receivers specially designed to detect ultrasound under water. They detect pressure variations of acoustic signals propagating in water and produce voltage proportional to the pressure. Important hydrophone parameters are frequency range, sensitivity, and noise level. They are operated below the resonance frequency and are required to have wide frequency range and high signal-to-noise ratio.

Projectors are ultrasonic transmitters used under water. They generate high-intensity ultrasonic waves when sinusoidal voltages are applied to them. The ultrasonic waves propagate in water up to long distances. The projectors are operated close to the resonance frequency. The output is a maximum at the resonance frequency.

4.7.1 Tonpilz Transducer

The Tonpilz transducer is a typical ultrasonic transducer used in SONAR applications. The transducer can be used as a projector or a hydrophone. The transducer uses either piezoelectric materials or magnetostrictive materials for actuation and sensing. A piezoelectric Tonpilz transducer consists of a stack of piezoelectric ceramic rings (PZT) with a tail mass of steel or brass and a head mass of a light metal such as aluminium or magnesium, all held together by a steel bolt as shown in Figure 4.17. The PZT rings are prestressed with compressive stress with the help of the steel bolt. A polymer coating is given to the entire system, which is enclosed in a waterproof package.

Important parameters of the transducer are its resonance frequency, transmission voltage response, receiving sensitivity, and electrical impedance. The size of the piezoelectric ceramic stack and the materials used for the tail mass and the head mass and their size and shape determine the resonance frequency, the bandwidth, and the transmitting and receiving sensitivities. The head mass of the transducer must be of light material like aluminium or magnesium in order to get increased bandwidth and also to get better acoustic impedance matching with water. The role of tail mass is to avoid backward acoustic radiation and it is made of heavy metal like steel or bronze. The size and shape of the tail mass are chosen to tune the resonance frequency of the system. Some designs include a suitable soft foam layer to be added to the end of the head mass to improve impedance matching with water. The piezoelectric rings are electroded on both sides, and alternate common faces are earthed. The other alternate faces are electrically connected, and voltage is applied to the faces.

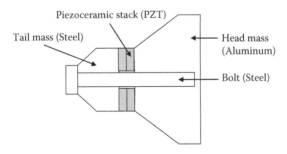

FIGURE 4.17
Schematic of a typical piezoelectric Tonpilz transducer.

The transducer can be used both as a transmitter and a receiver. When operated as a transmitter (projector), the transducer is excited at the resonance frequency by applying a sinusoidal or pulsed voltage to alternate faces of the piezoelectric discs in the stack. When used as a receiver (hydrophone), it is operated below the resonance frequency, and the voltage generated at the faces of the piezoelectric discs in the stack is measured after suitable amplification. Tonpilz transducers are usually designed to have resonance frequencies in the range of 10–50 kHz.

4.7.2 High-Frequency Underwater Transducers

Although Tonpilz transducers are widely used in SONAR applications, they have the limitation that their operating frequency cannot exceed 50 kHz. For higher-frequency operations, the size of the device will have to be much smaller, making the Tonpilz design unsuitable. High-frequency transducers are built with piezoelectric materials in the shape of rods, parallelepipeds, plates, or rings. The operating frequencies of the devices are determined by the thickness of the piezoelectric elements. Frequencies greater than 100 kHz can be achieved with these transducers.

An array of small piezoelectric elements may be used to increase the dimensions of the transducer. A linear array transducer consisting of an array of small piezoelectric blocks is schematically shown in Figure 4.18. The piezoelectric elements are mounted on a rigid backing material, and the entire assembly is enclosed in a plastic package or in a transparent fluid-filled case.

High-frequency transducers have been developed with composite piezoelectric materials. Composites of connectivity 0–3, 1–3, and 3–3 have been used. Piezoelectric ceramic PZT and polymers are used in the composite materials. Most of the composite piezoelectric transducers are used as hydrophones.

Commercial piezoelectric underwater transducers of various sizes and shapes manufactured by ITT, Electronic Systems are shown in Figure 4.19.

4.7.3 Piezoelectric Hydrophone Specifications

The sensitivity of a hydrophone, called *receiving sensitivity*, is expressed in dB relative to 1 V per micropascal (dB re 1V/μPa). The unit is a measure of the voltage generated per unit of sound pressure falling on the device. The

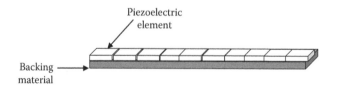

FIGURE 4.18
Linear array piezoelectric transducer.

FIGURE 4.19
Tonpilz transducers. Courtesy: ITT Electronic Systems-Acoustic Sensors.

voltage is given relative to a reference voltage equal to 1 V. The unit is given in decibels by

$$20\log\frac{V}{V_o} \text{ re 1 V per 1 } \mu\text{Pa}$$

V is the voltage output of the hydrophone for 1 μPa input pressure. The reference voltage V_o is equal to 1 V. The receiving sensitivity is normally expressed in negative values, for example −120 dB re 1 V per 1 μPa or −140 dB re 1 V per 1 μPa, etc.

Example:

If the receiving sensitivity of a hydrophone is −120 dB re 1 V per 1 μPa, then

$$20\log\frac{V}{V_o} = -120$$

$$\log\frac{V}{V_o} = -6; \quad \frac{V}{V_o} = 10^{-6}; \quad \text{i.e.,} V = 10^{-6} \text{ volts}$$

That is, the output of the hydrophone is 1 μV for an input sound pressure of 1 μPa.

4.7.4 Piezoelectric Projector Specifications

The sensitivity of the projector is called the Transmitting Voltage Response (TVR), and it is expressed in dB re 1 µPa per V at 1 m distance. The unit is a measure of the acoustic pressure relative to 1 µPa at a distance of 1 m from the transducer, for an input voltage of 1 V.

TVR is expressed in decibels as

$$\text{TVR} = 20\log \frac{P}{P_0} \ re \ 1\mu \ \text{Pa per } V \text{ at } 1 \text{ m}$$

where P is the pressure at a 1 m distance for a 1 V input to the projector. The reference pressure P_0 is equal to 1 µPa.

Example:

If the TVR of a projector is 150 dB re 1 µPa per 1 V, then

$$\text{TVR} = 20\log \frac{P}{P_0} = 150$$

$$\log \frac{P}{P_0} = 7.5; \quad \frac{P}{P_0} = 10^{7.5} = 31.62 \times 10^6; \quad \text{i.e., } P = 31.62 \times 10^6 \times 10^{-6} \text{ Pa} = 31.62 \text{ Pa}$$

That is, the pressure at a 1 m distance from the projector for an input voltage of 1 V is 31.62 Pa.

4.8 Piezoelectric Tactile Sensor

Tactile sensors are devices which sense contact or touch. The sensors have applications in industrial automation and robotics. In the medical field, they are useful in laparoscopic surgery. The sensor gives an output signal whenever it comes into contact with an object. It is a touch sensor just like human skin. The sensor can be used in a robot to make it manipulate delicate objects. It can also be used to determine the contour of objects. It is useful in noninvasive surgery in which the scalpel and grasper need to be carefully handled inside the human body.

There are different types of tactile sensors based on the principle used, for example, resistive, capacitive, optical, and piezoelectric. In piezoelectric tactile sensors, the direct piezoelectric effect is used where a small pressure on the sensor generates an electrical signal. More sensitive piezoelectric tactile sensors make use of the change in resonance frequency of a piezoelectric element when a small pressure acts on it.

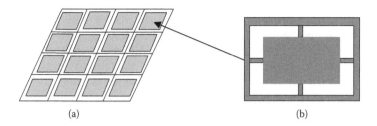

(a)   (b)

FIGURE 4.20
Schematic of a tactile sensor. (a) A 4×4 array tactile sensor. (b) Each element of the tactile sensor—suspended bridge type.

Piezoelectric materials used for the tactile sensor are thin films of ceramic PZT, polymer PVDF, or copolymer of PVDF [11–13]. It is necessary that the tactile sensor be flexible, and so polymers are better suited.

A piezoelectric tactile sensor consists of a large number of tiny sensors arranged in a matrix. Each element is a piezoelectric cantilever or bridge micro-machined on a silicon substrate. The electronics required is integrated on the same chip. A tactile sensor array is schematically shown in Figure 4.20a. Each element of the array consists of a suspended piezoelectric membrane as shown in Figure 4.20b.

The size of the tactile sensor depends on the application. The area of the array may range from 1 cm^2 to several cm^2 and the number of elements in the array could be 10 to 100 elements. The greater the number of elements, the better will be the resolution.

The output from each element could be binary output (touch and no-touch) or could be analog output voltage proportional to the pressure exerted on the element. In the binary type, the position of contact can be determined, and in the analog type both the position and the extent of pressure can be determined.

A tactile sensor in which the change in resonance frequency of the piezoelectric element is measured is more sensitive than the one in which the output voltage is measured [14].

4.9 Energy Harvesting

Energy harvesting using the piezoelectric effect is a relatively new technique of using piezoelectric materials as sources of electrical energy for operating low-power devices. The technique makes use of the direct piezoelectric effect, in which mechanical vibrations get converted to electrical signals. There are many common sources of mechanical vibrations and movements existing in our normal environment. For example, walking and other movements of humans and noise and vibrations that occur in normally used gadgets and machines are the sources that generate mechanical vibrations in our daily life. Movement of traffic in a busy street is another common source of noise and vibration which can be utilized for power generation. Many novel and new ideas are being proposed and published on this new alternative energy source. However, the amount of power generated is relatively low, and so the technique can be utilized only for low-power applications. The electrical energy generated may not necessarily be used directly but may be used to charge normal batteries, which are then used to operate various low-power devices. Examples of some extremely useful low-power devices that can be operated with this technique are mobile phones, electronic watches, iPods, hearing aids, and medical implants. The electrical output one can get out of these piezoelectric energy-harvesting devices depends on the size and mass of the device. The bigger the size and mass, the greater will be the output power. Since the size of the energy source needs to be very small for operating small devices, the output power is limited. Although the techniques of this type of energy harvesting have developed quite well, it would become a commonly used energy source only if the power requirements of electronic devices get reduced further.

The frequency range of normally occurring vibrations in our environment (body movements, electric gadgets, etc.) lie in the range of about 50–250 Hz. The piezoelectric energy-harvesting device must have resonance frequency within this range to give considerable electrical energy output. The smaller the device, the higher will be the resonance frequency. So the device cannot be made very small.

The mean power that can be harvested from a piezoelectric device subjected to random vibrations has been investigated theoretically by several investigators [15,16]. Among the various designs, the piezoelectric bimorph cantilever described in Section 4.11 has been found to be most suitable for energy harvesting and has been widely studied [17–19]. Piezoelectric cymbal transducers described in Section 4.11 have also been used for energy harvesting [20].

The bimorph or the cymbal transducer is mounted on the vibrating system. The vibration gets transferred to the bimorph, and it is forced to vibrate. Due to the direct piezoelectric effect, the vibrating bimorph generates an AC electrical voltage. The sensitivity of the device is increased when a proof mass is added at the free end of the bimorph as shown in Figure 4.21.

The bimorph has a natural frequency of resonance which is determined by its dimensions and the value of the proof mass. To have maximum efficiency, it is necessary that this natural frequency exists with high amplitude in the frequency band of the vibrating system. The AC voltage generated by the bimorph is rectified to get DC voltage. The efficiency will improve if the mechanical damping of the vibrating bimorph is minimized and the electromechanical coupling constant of the piezoelectric material is high.

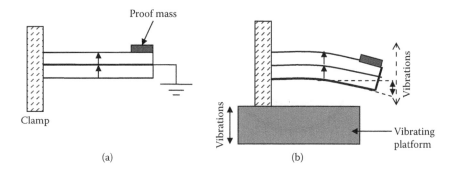

FIGURE 4.21
(a) Piezoelectric bimorph used for energy harvesting. (b) The vibrating platform transfers the vibrations to the bimorph. AC voltage is generated across the electrodes of the bimorph.

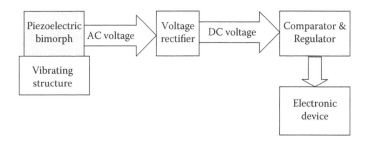

FIGURE 4.22
Block diagram of piezoelectric energy-harvesting system.

A block diagram of the energy-harvesting system is shown in Figure 4.22. The vibrating system could be any naturally vibrating structure in our environment, for example, a busy staircase in a building, a road with busy traffic, a dancing floor in a club, a vibrating engine inside an automobile, or any common electric gadget used at home such as a washing machine, kitchen blender, or refrigerator. A piezoelectric bimorph is the transducer that converts the mechanical energy of vibrations to electric AC voltage. The bimorph must be designed to have the natural frequency of vibration that coincides with the frequency of the highest amplitude in the frequency spectrum of the vibrating structure. The AC voltage output of the bimorph is converted to a DC signal using the rectifier circuit. The DC output is stored in a capacitor. The comparator and the regulator circuits limit the DC voltage to the voltage required for the operation of the electronic device.

Piezoelectric energy harvesting using the mechanical energy generated during walking by humans has been studied extensively [21,22]. This is done by embedding the piezoelectric device inside the soles of shoes. Polymer piezoelectric material PVDF, which is more flexible and less brittle, is better suited than the stiff and brittle piezoelectric ceramic (PZT) for such applications.

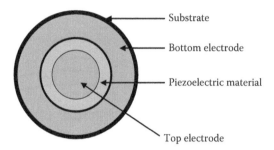

FIGURE 4.23
Piezoelectric diaphragm used in buzzers and tweeters. Electrical connections are taken from the top and the bottom electrodes of the piezoelectric material.

4.10 Piezoelectric Electronic Buzzer and Tweeter

Piezoelectric Electronic Buzzer and Tweeter are both piezoelectric devices which convert AC electrical signals to mechanical vibrations and give out sound. Piezoelectric buzzers operate at low frequencies, whereas tweeters are speakers that operate at relatively higher frequencies. Buzzers are used as beepers in watches and many electronic devices and tweeters which operate at high frequencies, typically in the range 2–20 kHz, are used as speakers in mobile phones, computers, and portable radios. Some of the advantages of piezoelectric speakers over conventional speakers are that they can be made extremely small in size, they do not get affected by magnetic fields, and they need much less electric power.

The physical structure of buzzers and tweeters is basically the same. It consists of a thin piezoelectric diaphragm which vibrates when electrical signal is applied across its thickness. The dimension of the diaphragm and the way it is mounted in the assembly determine the useful frequency range of the device. Piezoelectric materials used are quartz, barium titanate, or PZT. An electroded piezoelectric diaphragm used in these devices is schematically shown in Figure 4.23.

4.11 Piezoelectric Actuators

The indirect piezoelectric effect, in which an electrical input changes the dimensions of a piezoelectric material, has resulted in the development of many novel piezoelectric actuators for varied applications. The advantages of piezoelectric actuators are that they do not require complex designs, and they can generate considerable forces with quick response for low-voltage

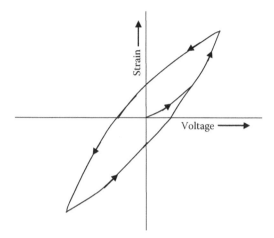

FIGURE 4.24
Strain versus voltage for a piezoelectric material. Hysteresis effect is exhibited by the materials.

input. A typical strain versus voltage curve of a piezoelectric material is shown in Figure 4.24. The materials exhibit a slight hysteresis effect, but this does not affect the actuator behaviour much if taken care of in the design.

Three types of commonly used piezoelectric actuators are

- Piezoelectric stack actuator
- Piezoelectric bender actuator
- Piezoelectric cymbal actuator

4.11.1 Piezoelectric Stack Actuator

Piezoelectric materials exhibit a strain of about 0.1% to 0.2% even for relatively high voltages. One of the methods for effectively using the material as a sensitive actuator is to use many piezoelectric elements in a stack. This enhances the strain for a given electric voltage.

A piezoelectric stack actuator consists of several thin piezoelectric discs or rings stacked one above the other as shown in Figure 4.25.

Adjacent discs are poled in opposite directions. The faces of each element are electroded. Alternate faces are electrically connected, and a voltage is applied across the electrodes as shown. The change in dimension of each of the elements gets added up. If there are N number of elements in the stack and each element changes its dimension by Δl, the total change in dimension of the stack would be $N\Delta l$.

For most applications, the size the actuator cannot be too big, and the dimensions will have to be within a few centimetres. The number of elements in the stack has to be limited, and the size of each element also must be small. For a stack of size about 10 mm length, the maximum stroke that

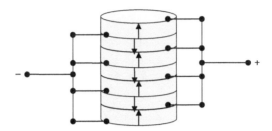

FIGURE 4.25
Piezoelectric stack actuator. Adjacent discs are poled in opposite directions. Alternate faces are electrically connected.

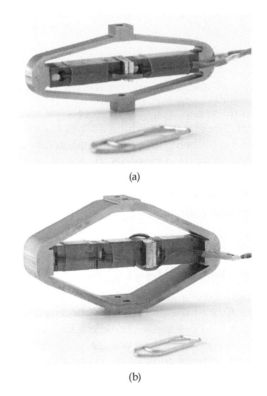

(a)

(b)

FIGURE 4.26
A piezoelectric stack actuator with mechanical amplification. (a) Stack actuator. (b) Magnified lateral displacement. (Courtesy: CEDRAT Technologies.) The expansion of the stack creates a lateral amplified movement along the minor axis of the elliptic shell.

could be obtained is about a fraction of a micrometer. To improve the stroke, the actuator displacement is magnified using mechanical designs. One such design is shown in Figure 4.26 (Courtesy CEDRAT Technologies, France). The piezoelectric actuator consists of rectangular piezoelectric elements. The stack is mounted along the axis of an elliptical stainless steel shell as shown in Figure 4.26. The stack expands or contracts along the axis when an external voltage is applied across the electrodes. This change in dimension generates a magnified lateral displacement of the elliptical shell along the minor axis. The displacement is used as the stroke of the actuator.

Piezoelectric stack actuators have applications as valves, switches, relays, micro-positioning device, shutters, and optical aligning devices. They are used in smart structures, robots, etc.

4.11.2 Piezoelectric Bender Actuator—Bimorph and Unimorph

Two types of piezoelectric bender actuators are piezoelectric bimorph and piezoelectric unimorph.

4.11.2.1 Piezoelectric Bimorph

The piezoelectric bimorph consists of two thin strips of piezoelectric elements bonded to each other. The bimorph can be used in two configurations: (1) series and (2) parallel. The bimorph is clamped at one end, and the other end is free.

In a series configuration, the two strips are bonded such that the two are poled in opposite directions (Figure 4.27a). The outer faces of the strips are electroded, and a voltage is applied across the two electrode faces as shown. The bias across each of the strips is in the same direction, but the two are poled in opposite directions. So the upper strip expands,

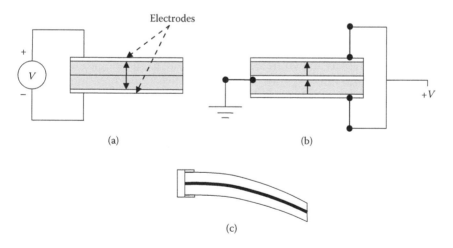

(a)   (b)

(c)

FIGURE 4.27
Piezoelectric bimorph. (a) Series configuration: The two strips are oppositely poled. Voltage is applied across the two outer electrodes. (b) Parallel configuration: The two strips are poled in the same direction. Voltage is applied between the outer electrode and the middle electrode. (c) Bimorph in the bent form.

and the bottom strip contracts due to the transverse piezoelectric effect (d_{31} coefficient is involved). This makes the bimorph bend downwards (Figure 4.27c).

In a parallel configuration, the two strips are bonded such that the two are poled in the same direction (Figure 4.27b). The common face of the strips and the two outer faces are electroded. The common face is grounded, and a positive voltage is applied to the two outer electrodes. The upper strip expands and the bottom strips contracts, resulting in bending of the bimorph (Figure 4.27c).

The piezoelectric bimorph is capable of generating relatively high displacements. Applications of bimorphs are in precision positioning, valves, switches, etc. With the advent of microelectromechanical systems (MEMS), micro-fabrication techniques have been used to fabricate microbimorphs which have applications in micro-motors and micro-valves in engineering and medical fields.

4.11.2.2 Piezoelectric Unimorph

The piezoelectric unimorph is a cantilever consisting of a single strip of piezoelectric material clamped at one end and free at the other end. The bending of the cantilever on application of voltage is used for actuation. The unimorph is made of a thin film of piezoelectric material formed on a non-piezoelectric substrate material. The system is clamped at one end, and a voltage is applied across the piezoelectric film. Because of the piezoelectric effect, the piezoelectric film gets strained. The substrate, which is non-piezoelectric, resists the strain. This results in bending of the unimorph.

The unimorph can be used in two configurations: d_{33} mode and d_{31} mode.

In d_{33} mode, the piezoelectric layer is poled in a direction parallel to the plane (direction 3) as shown in Figure 4.28a. Due to the longitudinal piezoelectric effect, the film expands in the same direction (direction 3), which results in bending of the unimorph.

In d_{31} mode, the piezoelectric layer is poled in a direction perpendicular to the plane (direction 3) as shown in Figure 4.28b. Due to the transverse piezoelectric effect, the piezoelectric film expands in the transverse direction, which is direction 1. This results in bending of the unimorph.

Using MEMS micromachining techniques, micron-size piezoelectric unimorph actuators have been fabricated. The piezoelectric films are deposited on silicon oxide or silicon nitride substrates. Piezoelectric materials used are PZT, ZnO, and polymer PVDF. The surface micromachining technique is used for fabrication of the cantilever structure. A schematic diagram of a micromachined piezoelectric unimorph actuator is shown in Figure 4.29.

Piezoelectric bimorph and unimorph are used as rf switches, atomic microscope tips, relays, etc.

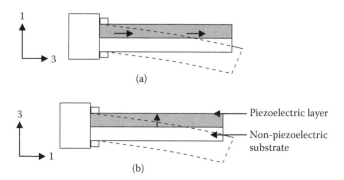

(a)

(b)

FIGURE 4.28

Piezoelectric unimorph. (a) d_{33} mode: the piezoelectric layer is poled in direction 3, and the material is strained in the same direction (direction 3). (b) d_{31} mode: the piezoelectric layer is poled in direction 3, and the material is strained in direction 1.

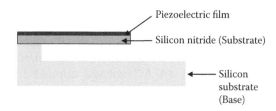

FIGURE 4.29

Schematic of a micromachined piezoelectric unimorph actuator.

4.11.3 Piezoelectric Cymbal Actuator

The cymbal actuator, which was first developed by the Newnham group at Penn State University [23], exhibits high sensitivity because of its unique design. It basically consists of one or two circular piezoelectric ceramic discs sandwiched between two cymbal-shaped metal end caps. Schematic diagrams of single-disc and double-disc cymbal actuators are shown in Figures 4.30a and 4.30b.

In the single-disc cymbal transducer, the PZT disc is electroded on both sides and is poled in the thickness direction. Voltage is applied across the electrodes. In the double-disc cymbal transducer, the two discs are poled in opposite directions. Voltage may be applied to the common face with the outer two faces earthed, or voltage may be applied to the two outer faces with the common face earthed.

The end caps, made of thin stiff metal (titanium, brass, or steel), are bonded to the PZT disc on both sides using epoxy glue. The metal caps act like a

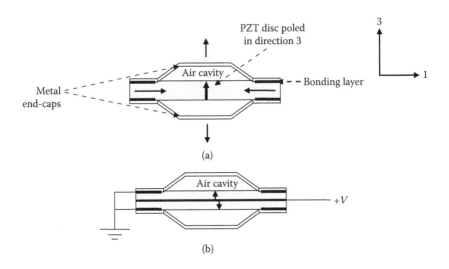

FIGURE 4.30
Piezoelectric cymbal transducer: (a) single disc, (b) double disc.

mechanical transformer which transforms the small extensional motion of the piezoelectric ceramic disc into a large flexural motion of the metal caps.

When used as an actuator, an AC voltage applied to the PZT disc across its thickness causes the disc to change its dimension in direction 3 (d_{33} coefficient). This causes lateral contraction along the radial direction (d_{13} coefficient), resulting in an amplified displacement of the metal caps. Thus, both d_{33} and d_{13} coefficients contribute to the actuator action. The choice of the material for the metal caps and the cavity depth determine the sensitivity of the actuator. The sensitivity of a cymbal actuator is higher than a bimorph or a stack actuator.

The device can also be used as a sensor. When used as a sensor, the vibration or displacement of the end caps causes a voltage to be developed across the piezoelectric disc which is measured.

The resonance frequency of the transducer can be varied over a wide range by suitably selecting the diameter of the piezoelectric disc and the cavity depth of the end caps. Cymbal transducers with PZT discs of varied diameters in the range of 3–50 mm and with end cap cavity depths in the range of 0.2 to 0.5 mm have been designed and reported in the literature. The resonance frequencies of the transducers lie in the range of 1–100 kHz [24,25]. High-frequency cymbal transducers are used as ultrasonic generators and detectors in underwater acoustics. Cymbal transducers may be used as single elements or may be assembled into large flexible arrays. Array transducers with cymbal transducer elements in two-dimensional arrays have been designed as underwater projectors [26]. The array can be designed to have quite a wide range of frequency response.

Applications of cymbal transducers include accelerometers, micro-positioners, switching elements for valves, vibration sensing and suppression in automobiles, ultrasonic generators (projectors) and sensors (hydrophones) in underwater acoustics, energy harvesting, etc. In the medical field, cymbal transducers have been widely studied and applied as an ultrasonic source for transdermal drug delivery [27,28].

4.12 Piezoelectric Motor

A motor is a device that produces continuous linear or rotary motion. Piezoelectric motors, which make use of the indirect piezoelectric effect to produce motion, have several advantages over conventional electromagnetic motors. They do not need strong magnetic fields as conventional electromagnetic motors do. Other advantages of piezoelectric motors are that they can be miniaturized, can be operated at much lower power, and are more reliable. The micro-motors satisfy the requirements of precise positioning applications such as mask alignment in IC technology, fibre optic alignment, medical catheter placement, auto-focus and optical zoom in mobile phone cameras, pharmaceuticals handling, etc. They can be miniaturized to a size of less than 4 mm and can provide position accuracies up to 0.1 micrometers or better. They have many other applications in both engineering and medical fields wherever precision movements are required and where a magnetic field in the vicinity is to be avoided. Two types of piezomotors are linear motor and rotary motor. In these motors, the small displacement of a piezoelectric material due to the indirect piezoelectric effect is converted into continuous translatory motion or a rotary motion.

4.12.1 Linear Piezoelectric Motors

Linear piezoelectric motors are of two types: one type operates at low frequencies and the other at ultrasonic frequencies.

One type of low-frequency linear piezoelectric motor is the clamping type, which produces linear motion of a slider just like the movement of a worm [29]. This type of piezomotor, called the "inch worm" motor, was first designed and patented by Burleigh Instruments Inc. The principle of the motor can be understood from Figure 4.31.

The inch worm motor consists of three sets of piezoelectric actuators: two are clamping actuators (1 and 2), and one is the driving actuator (3), as shown in Figure 4.31. The actuators are operated in sequence to achieve motion of the slider, which is to be moved.

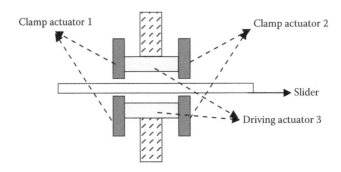

FIGURE 4.31
Schematic of a piezoelectric inch worm motor.

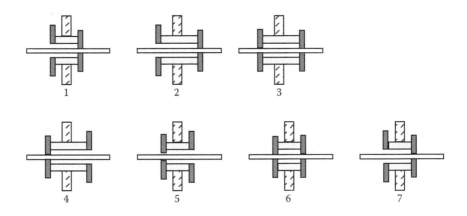

FIGURE 4.32
The seven steps (one cycle) of operation of an inch worm motor.

The two actuators 1 and 2 are the clamp actuators, which hold the slider alternately. Actuator 3 is the drive actuator, which gets extended or contracted in the direction of the slider movement. Initially, all the actuators are disconnected from the slider. When one of the clamp actuators clutches the slider, the other clamp actuator is disconnected. Alternately, one of the clamp actuators clutches the slider. The drive actuator extends or contracts laterally in the direction of the movement of the slider. The steps of operation as illustrated in Figure 4.32 are as follows:

Initially, all the actuators are open and inactive.

Step 1: Clamp actuator 2 is closed.

Step 2: The drive actuator gets extended.

Step 3: Clamp actuator 1 is closed.

Step 4: Clamp actuator 2 is opened.

Step 5: The drive actuator gets contracted.

Step 6: Clamp actuator 2 is closed.

Step 7: Clamp actuator 1 is open.

These steps get repeated to operate the linear motor.

4.12.2 Piezoelectric Ultrasonic Rotary Motor

Piezoelectric ultrasonic rotary motors are of two types: (1) standing wave type and (2) travelling wave type.

The standing-wave-type piezoelectric motor works on the principle of generating a standing wave in a piezoelectric ring which acts as a stator. In one of the designs [30], a piezoelectric ring vibrating radially at its resonance frequency transfers the vibrations to a set of spring pushers which are attached to the rotor as shown schematically in Figure 4.33. The pushers are so arranged that the radial vibration of the ring generates perpendicular vibration of the pushers because of the phase shift. This makes the pushers move in an elliptical path. The motion is transferred through friction to the rotor causing rotation of the rotor. The motor can be operated in a continuous mode or stepper mode. The resonance frequency of the ring is in the ultrasonic range (50–80 kHz). For continuous mode, the sinusoidal voltage of the required frequency is applied to the ring. For stepper mode, the voltage at the US frequency is gated to get pulses of fixed pulse frequency and duty cycle.

The travelling-wave-type piezoelectric motor consists of a piezoelectric ring vibrating at its natural frequency. The stator, which is a thin annular metallic ring, is attached to the piezoelectric ring as shown in Figure 4.34. The rotor with a shaft is connected to the stator. The stator and the rotor have a common shaft with a bearing at the centre as shown. The piezoelectric ring is excited at its natural frequency (30–50 kHz) to generate a travelling wave.

FIGURE 4.33
Schematic of a piezoelectric ultrasonic motor—standing wave type.

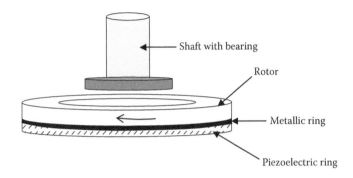

FIGURE 4.34
Schematic of a piezoelectric travelling wave rotary motor.

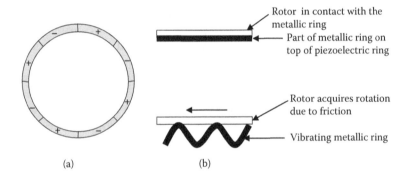

FIGURE 4.35
(a) Polarization of the piezoelectric ring. (b) Operation of the motor: Top: Rotor in contact with the metallic ring. Bottom: Metallic ring subjected to vibration and translation of the vibration to rotation of the stator through friction.

For generation of the travelling wave, the piezoelectric ring is divided into n sections, and each section is polarized such that alternate sections are polarized in the same direction, and adjacent sections are polarized in the opposite direction, as shown in Figure 4.35a. Sinusoidal voltages are applied to the two groups such that they get voltages 90° out of phase. A travelling wave is thus set up in the piezoelectric ring, and the stator follows the vibrations of the piezoelectric ring (Figure 4.35b). The rotor in contact with the stator acquires rotational motion through friction.

The resonance frequency is determined by the size of the motor. The motors can be configured with dimensions in the range of 20–60 mm.

Piezoelectric motors of several designs have been designed and patented by many research workers [31–33]. Extremely small piezoelectric motors of a few millimetres in size have been made using micro-fabrication techniques (MEMS).

4.13 Piezoelectric Micro Pump

Micro pumps are essential components of microfluidic systems, in which extremely minute volumes of fluids are required to be transported in a controlled manner. Examples of microfluidic systems include inkjet printers, fuel injection systems, gas sensors, biochemical assays, genetic analysis system, drug delivery systems, etc. Several actuation mechanisms have been used in the design of micro pumps. Piezoelectric micro pumps use piezoelectric actuators for generating the driving force.

A piezoelectric reciprocating displacement pump basically consists of a pump chamber with a diaphragm, two valves, and an actuator to drive the pump diaphragm. In MEMS-based micro pumps, the chamber is formed in a silicon substrate sandwiched between glass plates using micromachining techniques. The chamber is provided with two one-way valves for input and output. A deformable plate made of silicon, glass, or plastic which acts as the diaphragm is formed on one side of the chamber. A piezoelectric actuator is used to actuate the diaphragm. In practice, several layers of materials are used to design the micro pump. The operation is schematically illustrated in Figure 4.36.

The piezoelectric actuator drives the diaphragm to alternately increase and decrease the pump chamber volume. In the suction stroke (Figure 4.36a), the chamber volume is increased. The fluid from the reservoir is sucked into the chamber through the inlet valve. In the discharge stroke (Figure 4.36b), the chamber volume is decreased. The fluid in the chamber gets discharged out through the outlet valve.

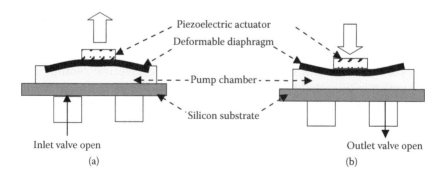

FIGURE 4.36
Schematic illustration of the operation of a reciprocating displacement micro pump using a piezoelectric actuator: (a) suction stroke, (b) discharge stroke.

Reciprocating displacement pumps with more than one chamber and with as many actuators have been designed. The actuators are operated in succession to make the fluid flow from the inlet to the outlet. A MEMS-based piezoelectric micro pump with three chambers is reported by J. G. Smits [34]. The micro pump operates at frequency of 15 Hz and is capable of a pumping rate of 0.008 ml/min. Piezoelectric micro pumps of various designs capable of pumping rates ranging from a fraction of a millilitre per minute to a few millilitres per minute have been reported in the literature. The operating frequencies of the various pumps are in the range of 15–150 Hz. Piezoelectric materials used for actuation are PZT or polymeric piezoelectric PVDF. Reciprocating displacement pumps are useful where continuous flow of controlled amounts of fluid is required.

Another type of piezoelectric micro pump is the drop-on-demand type, which is used in ink-jet printers. The pump uses a piezoelectric actuator to draw a small amount of ink from the reservoir and force it out of the nozzle in the form of small droplets. The input voltage pulses to the actuator can be programmed to get a precisely controlled amount of ink from the nozzle at the required time.

4.14 Piezoelectric Ultrasonic Drill

The indirect piezoelectric effect has been effectively used for the design of a piezoelectric drill. The vibration of the piezoelectric material at high ultrasonic frequencies is utilized to actuate the tip of the drill. High-frequency vibrations of the tip can drill precision holes through hard materials. Deep hole drilling of extremely small diameters of the order of 1 mm or less has been achieved with piezoelectric drills. Other advantages of a piezoelectric ultrasonic drill are high surface quality, capability of achieving different shaped holes, and capability of drilling holes in ceramic materials.

The piezoelectric drill consists of a stack of piezoelectric ceramic rings (PZT) which is subjected to vibrations at ultrasonic frequencies (20–50 kHz) by exciting with an AC voltage. The longitudinal vibration of the piezoelectric material is transferred to the drill bit, which is suitably fixed to the ceramic stack. The piezoelectric drill is specially suited in applications where small, lightweight, portable drills are required. Examples are in dental surgery, orthopaedic surgery, and in space exploration.

Dr. Yoseph Bar-Cohen and his team [35,36] have reported the design of a piezoelectric drill used for NASA exploration of extraterrestrial bodies. The drill has been used in Mars exploration to drill core samples of rocks on Mars for testing its constituents [37,38]. The detailed structure of the drill and results of a finite element analysis of the drill carried out by the author using PAFEC software tools are described in Chapter 6 of the book (Section 6.5.2).

4.15 Ultrasonic Cleaner

Ultrasonic cleaners use piezoelectric or magnetostrictive ultrasonic generators for producing ultrasonic fields. Most of the commercially available ultrasonic cleaners use piezoelectric generators. The frequency of operation of ultrasonic cleaners is in the range of 20–40 kHz.

In an ultrasonic cleaner, an ultrasonic field produces high-frequency pressure waves in a suitable liquid medium. The objects to be cleaned are kept immersed in the liquid. High-frequency pressure waves generate millions of micron-size vacuum bubbles in the liquid, which continuously keep collapsing and reappearing. The phenomenon of formation and collapse of micro-bubbles due to high frequency pressure waves is called *cavitation*. When the high-frequency pressure waves travel in the liquid medium, tiny bubbles are created at the sites of rarefaction; the size of these bubbles oscillate due to the periodic pressure variation and when they eventually grow to an unstable size, they collapse violently, resulting in implosions. This causes high temperature and pressure to be generated at the sites. A large number of bubbles constantly growing and collapsing in the liquid medium helps in displacing the contaminants from the surface and other small cavities in the object to be cleaned. The liquid used in the cleaner must be a suitable solution that can dissolve the contaminants thrown out of the object.

Ultrasonic cleaning enables cleaning of interior surfaces of complex machinery parts with a high degree of cleaning levels. They are used for cleaning objects such as substrates for thin-film coating, small complex parts of industrial equipments, surgical instruments, jewellery, etc.

A piezoelectric ultrasonic cleaner consists of a stack of piezoelectric discs with a steel back-mass and a front aluminium coupling mass assembled under the bottom of a strong steel tank. Cleaning liquid solution is filled in the tank, and the objects to be cleaned are kept immersed in the liquid. The schematic diagram of a piezoelectric ultrasonic cleaner is shown in Figure 4.37. The piezoelectric stack consists of a few piezoelectric circular discs with a central hole. A compression bolt is used to compress the stack between a steel back-mass and front coupling mass as shown. A steel plate fixed on top of the front mass is attached to the bottom of the steel tank, which contains the cleaning liquid.

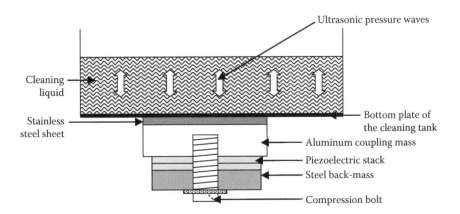

FIGURE 4.37
Schematic of an ultrasonic cleaner.

The cleaning liquid is usually an aqueous solution of a suitable detergent which is able to dislodge the contaminants effectively. The solution must be chosen according to the type of contaminants to be removed from the object. Another important consideration in choosing the solution is that it must have high vapour pressure and low surface tension for an efficient cavitation effect. The temperature of the liquid is normally maintained to be in the range of 50°–60°C.

Ultrasonic cleaners are made in different sizes ranging from table-top types used for cleaning small objects such as jewellery, thin film substrates, small machinery parts, surgical instruments, etc., to very large-sized units for cleaning big industrial equipment parts.

4.16 Quartz Crystal Oscillator

A quartz crystal oscillator is a device that makes use of both direct and indirect piezoelectric effects. The AT-cut thin crystal in the form of a thin plate or in the form of a tuning fork is used for the crystal oscillator.

The device is used as a component in an electronic oscillating circuit. The output electrical signal is fed back to the crystal by which the oscillation can be maintained. This results in a stable frequency output signal. Stable frequency oscillators have many applications; for example, in transmitters and receivers for communication, in computers and other computer-controlled devices for precise clock signals, and for keeping precise time in clocks and watches.

The AT-cut quartz plate is electroded by depositing gold on its faces and is used as a component of an electronic feedback amplifier circuit as shown in Figure 4.38.

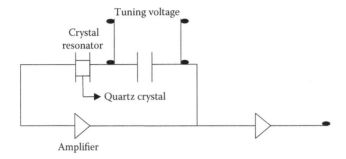

FIGURE 4.38
Quartz oscillator circuit.

The oscillator is tuned to the natural frequency of the quartz crystal. The small amount of energy fed back to the crystal causes it to vibrate at its natural frequency. The vibrations produce electrical signal at the same frequency, which is fed back to sustain the vibration at the stabilized frequency.

The resonance frequency of the quartz crystal used in watches and clocks is selected as 32.768 kHz. This frequency satisfies the requirements of suitable crystal size, battery life, and stability. The signal from the oscillator is converted to electrical pulses of the required frequency (one per second) using electronic circuits. The electrical pulses are used to drive stepper motors (in watches or clocks with hands) or to drive seven segment display devices (for digital watches or clocks).

In applications such as computer clocks and communication transmitters and receivers, crystals of much higher resonance frequencies, of the order of several megahertz and higher, are used.

4.17 Quartz Crystal Balance

A quartz crystal balance is based on the principle that the frequency of a vibrating quartz crystal gets shifted when a small mass is added to the electrodes of the oscillating crystal plate. Since a very small frequency shift can be accurately measured using electronic circuits, the sensitivity of the quartz crystal balance is inherently very high. Weights of the order of nanograms and less can be measured with the balance.

AT-cut quartz crystals are made to oscillate in shear mode at the resonance frequency by applying AC voltage. The frequency is normally in the range of several megahertz. Quartz crystal tuning forks have also been used for micro balance. Normally, the balance is used in vacuum or in gaseous environments.

Applications of such sensitive balances are in biomedical sciences, electrochemistry, and tribology.

4.18 Quartz Tuning Fork in Atomic Force Microscope

In an atomic force microscope (AFM), a vibrating quartz tuning fork is used to detect the atomic force between the probe tip and the scanned surface. A tuning fork of extremely small dimensions (e.g., length × thickness × width = $1500 \times 100 \times 100$ μm) is fabricated using a micromachining technique. The fork is integrated with the probe tip of the AFM. As the probe tip scans the surface of the specimen, the force acting on the tip causes the resonance frequency of the tuning fork to change, which is detected and stored in the memory as a signal. Thus, the contour of the surface on an atomic scale can be obtained as stored data in the memory which is converted to an image.

4.19 Piezoelectric Transformer

A piezoelectric transformer is a device used for stepping up or stepping down an input voltage. Both direct and indirect piezoelectric effects are used in the transformer. The advantages of piezoelectric transformers over conventional electromagnetic transformers are that they can be made very small in size, are of light weight, and can be used over a wide range of high voltages. They have applications wherever high voltage is required with limitations of space. The applications of step-up transformers include power supply for cold cathode fluorescent lamps for operating LCD in laptops, sub-notebook computers, flat screen televisions, and other Internet appliances. Step-down transformers are used for adapters and converters.

Three types of piezoelectric transformers are

- Rosen piezoelectric transformer
- Longitudinal vibration transformer
- Radial vibration transformer

The first two types are more widely studied and are used in many applications.

4.19.1 Rosen Piezoelectric Transformer

The Rosen piezoelectric transformer, which was first designed and patented by C. A. Rosen in 1961, makes use of both longitudinal and transverse modes of vibrations. The transformer consists of two thin piezoelectric plates attached adjacent to each other as shown in Figure 4.39. The two piezoelectric plates are poled differently; plate A, called the primary, is poled along the thickness direction (direction 1) and has electrodes on the top and bottom,

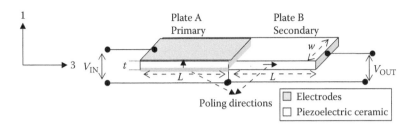

FIGURE 4.39
Rosen piezoelectric transformer.

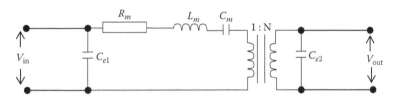

FIGURE 4.40
Equivalent circuit of Rosen piezoelectric transformer.

while plate B, called the secondary, is poled along the length direction (direction 3) and has electrodes on the edge faces as shown.

The input voltage is given to the primary across the electrodes (across the thickness of the plate—direction 1). The transverse indirect piezoelectric effect (d_{31}) causes the plate to vibrate in the perpendicular direction 3. The vibration of the primary plate gets mechanically transferred to the secondary plate, causing the plate to vibrate in direction 3, along which it is poled. This vibration causes a voltage to be developed across the electrodes in direction 3 due to the longitudinal direct piezoelectric effect (d_{33}). Thus, the input voltage to the primary plate gets transferred through mechanical vibrations to an output voltage across the secondary plate. The transformer is operated at its resonance frequency, which is usually in the range of 50–60 kHz.

The equivalent circuit of the Rosen transformer is shown in Figure 4.40.

C_{e1} and C_{e2} are the electrical capacitances of the primary and the secondary piezoelectric elements, respectively, given by

$$C_{e1} = \frac{\varepsilon L w}{t} \text{ and } C_{e2} = \frac{\varepsilon t w}{L}$$

where ε is the permittivity of the piezoelectric ceramic and L, w, and t are the length, width, and thickness of the two elements (Figure 4.39). R_m, L_m, and C_m are the mechanical resistance, mass (analogous to inductor), and compliance (analogous to capacitor), respectively. The values of R_m, L_m, and C_m depend

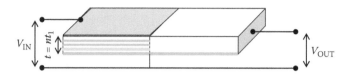

FIGURE 4.41
Multilayer Rosen piezoelectric transformer.

on the dimensions of the element. The multiplication factor N depends on the piezoelectric coefficients and also on the dimensions of the transformer.

$$N \propto \frac{L}{t}$$

If the transformer elements are longer and thinner, the output voltage will be higher. But a longer piezoelectric transformer will have a lower resonance frequency. The multiplication factor can be increased by using a multi-layer piezoelectric structure on the input side of the transformer as shown in Figure 4.41. In the multilayer transformer, the multiplication factor will be proportional to $(L/t_1)n$, where t_1 is the thickness of each layer and n is the number of layers.

The voltage gain (ratio of output to input voltage) of a typical multilayer Rosen type PZT transformer is in the range of 50–250 depending the load. The transformers are suitable for high-voltage applications, and they are used as step-up transformers. They are mostly used for fluorescent lamps for LCD displays in mobile electronic appliances, high-voltage sources for laser tubes used in copy machines, etc.

4.19.2 Longitudinal Vibration Piezoelectric Transformer

In the longitudinal vibration transformer, both the input and output piezoelectric elements vibrate in the longitudinal mode (thickness mode) [39]. The schematic of the structure is shown in Figure 4.42.

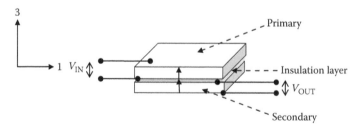

FIGURE 4.42
Longitudinal vibration piezoelectric transformer.

Two electroded piezoelectric plates are arranged one on top of the other as shown. They are separated by an insulation layer. Both the plates are poled in the thickness direction (direction 3).

The input voltage is given to the primary across the electrodes (i.e., across the thickness of the plate—direction 3). Longitudinal indirect piezoelectric effect (d_{33}) causes the plate to vibrate in the perpendicular direction 3. The vibration of the primary plate gets mechanically transferred to the secondary plate causing the plate to vibrate in the same direction (direction 3). This vibration causes a voltage to be developed across the electrodes in direction 3 due to longitudinal direct piezoelectric effect (d_{33}). Thus, the input voltage to the primary plate gets transferred through mechanical vibrations to an output voltage across the secondary plate. The multiplication ratio N of the transformer is proportional to the ratio of the thicknesses of the two plates. The transformer can also be made of multilayer piezoelectric elements. Both the primary and the secondary consist of thin layers of piezoelectric elements; if the primary plate has n_1 number of layers and the secondary plate has n_2 number of layers, the multiplication factor of the transformer N is proportional to the ratio (n_1/n_2).

Longitudinal mode transformers are used for low-voltage applications and mostly used as step-down transformers. Their applications include converters and adapters.

4.19.3 Radial Vibration Piezoelectric Transformer

The radial vibration piezoelectric transformer consists of two circular piezoelectric discs fixed one on top of the other as shown in Figure 4.43. The faces of the two discs are electroded, and they are poled along their thickness direction. The input voltage is given to the bottom disc, which acts as the primary element. The vibration of the disc due to the indirect piezoelectric effect is mechanically transferred to the secondary element (top disc). The radial vibration mode of the secondary disc is made use of, and hence it is called the radiation vibration transformer. The voltage generated due to the radial vibration of the secondary element (top disc) is measured as the

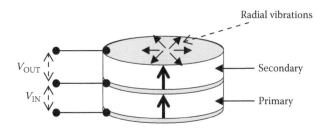

FIGURE 4.43
Radial vibration piezoelectric transformer.

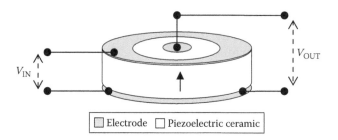

FIGURE 4.44
Single-disc radial vibration piezoelectric transformer.

output voltage. The wavelength of the fundamental vibration mode of the disc is given by $\lambda = 2r$, where r is the radius of the disc.

A radial piezoelectric transformer of output voltage 120 V and power 32 W which can be used as source for fluorescent lamp has been designed and reported by W. Huang [40].

A radial piezoelectric transformer using a single piezoelectric disc has been reported by P. Laoratanakul [41]. The transformer consists of a single piezoelectric disc electroded and poled in such a way that the input voltage causes vibration of the structure in the thickness mode, and the output voltage is due to the radial vibration of the same disc. The structure of the transformer is shown in Figure 4.44.

Two concentric circular electrodes are formed on the top face of the disc as shown, and the bottom face is fully electroded. An input voltage is applied between the top outer electrode and the bottom electrode, which is grounded. The radial vibration of the disc causes a voltage to be developed across the inner electrode of the top disc and the bottom grounded electrode which is measured as the output voltage of the transformer. The frequency of the input voltage is matched with the resonance frequency of the disc to get a large output voltage. With a piezoelectric ceramic disc of diameter 23.5 mm and thickness 1 mm, Laoratanakul has reported an output to input ratio of about 11 for a 10 kΩ load resistance. The power of the transformer is in the range of 1–7 W for input voltages in the range 5–35 V.

4.20 Nondestructive Testing (NDT) of Materials Using Ultrasonics

The most widely used technique for flaw detection and thickness measurement of materials is the Ultrasonic NDT technique. Piezoelectric transducers are most commonly used for the technique.

The principle of ultrasonic testing is transmission of high-frequency sound waves through the material and study of the propagation of the beam in the material through transmission, reflection, and attenuation. The study helps in detection of flaws such as cracks, porosity, corrosion, and impurities in the material. The technique can also be used to measure the thickness of the test material.

Piezoelectric transducers are used to generate ultrasonic waves. When an ultrasonic beam is incident on a solid material, both longitudinal and transverse waves propagate in the material. Longitudinal waves are those in which the particles in the medium vibrate parallel to the vibration direction. Transverse waves are those in which the vibration of the particles in the medium is perpendicular to the propagation direction. Longitudinal waves have higher velocity compared to transverse waves.

Two types of techniques used in NDT are (1) straight beam technique and (2) angle beam technique.

In the straight beam technique, a single transducer or two transducers may be used. In the single transducer method, the same transducer is used both as a generator and detector of ultrasonic waves. The ultrasonic waves in the form of pulses are generated at the transducer and allowed to propagate through the test material. The longitudinal ultrasonic beam passes straight through the material, and when it reaches the end face of the test object, gets reflected back. The reflected pulse is detected by the same transducer. Both the input pulse and the reflected pulse are displayed on a CRT screen. The separation between the two pulses gives information about the thickness of the sample. If the material has any flaws such as cracks, holes, or impurities in the path of the ultrasonic beam, additional pulses are observed between the two pulses because the waves get reflected at the flaw sites. The reflected pulses are of smaller amplitude because of absorption in the material. By accurately measuring the time interval between the pulses, the thickness of the sample and the location of the flaw can be determined. The straight beam ultrasonic transducer technique using a single transducer and the CRT display is illustrated in Figures 4.45a and 4.45b.

In the straight beam technique, two transducers may be used, one for generating the ultrasonic beam and the other for detecting the beam. The two are placed on either side of the test material. The ultrasonic beam generated by the generator is allowed to pass through the material, and the transmitted beam is detected at the other end by the detector. The two pulses, that is, the incident pulse and the transmitted pulse, are displayed on the CRT screen, from which the thickness of the material or presence of flaws in the path of the beam can be determined.

In the angle beam technique, two transducers are used. The ultrasonic beam is made incident on the material at an angle using a wedge-shaped plastic material at the head of the transducer. The incident transverse waves through the test material get reflected at the flaws or at the other end of

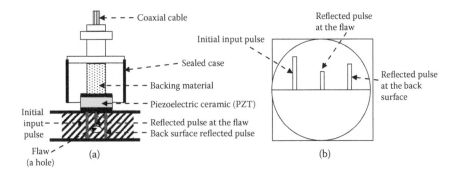

FIGURE 4.45
(a) Schematic of a straight beam piezoelectric NDT technique, (b) the input and output pulses from the device.

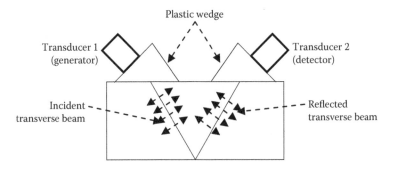

FIGURE 4.46
Schematic of an angle beam piezoelectric NDT device.

the material and reach the detector placed at a suitable position at an angle (Figure 4.46).

The probe angles of the transducers are so selected as to obtain the reflected beam from the sites where a flaw is suspected.

Piezoelectric ultrasonic transducers can be used as thickness gauges or flaw detectors for almost any materials: metals, plastics, or ceramics. In metals and ceramics, the attenuation is not much, whereas in plastics there is too much attenuation because of absorption. Therefore, the thickness range of plastic materials that can be used in this technique is limited.

4.21 Noise and Vibration Control

Vibrating structures are sources of noise, and it is important to control the vibrations not only to reduce the noise but also to protect the vibrating structures from damage. Thus, noise and vibration control is a very important scientific area which has received a lot of attention. It is an essential aspect of the design of equipment and machinery in civil engineering, automotive and aerospace engineering, and household equipment manufacturing. In precision equipments, wherever micro-positioning is required, such as read/write head in computers and CD players, mask alignment in integrated technology, etc., precise vibration control is of primary importance.

Two main types of vibration control are passive control and active control. Passive control is relatively simple as it uses vibration damping techniques such as the use of sound-absorbing materials or tuned dampers which use inertial mass and suitable viscous dampers. Active control is more complicated as it uses actuators and sensors to produce counter-vibrations to suppress or cancel the main vibrations.

Two techniques used for active vibration control using piezoelectric transducers are (1) shunt damping and (2) active feedback damping.

In shunt damping, a piezoelectric sensor is bonded to the vibrating structure, and the sensor is shunted by a passive electric circuit which can effectively dissipate the mechanical energy of the vibrating structure. A suitable RL circuit which is tuned to the required frequency is shunted across the electrodes of the piezoelectric sensor. The schematic diagram of the arrangement is shown in Figure 4.47.

The piezoelectric sensor is firmly bonded to the vibrating structure as shown. The sensor element may be a thin film of PZT or polymer PVDF. The sensor is shunted by a suitable electrical impedance consisting of a resistor and an inductor. When the structure vibrates, the AC voltage produced across the electrodes of the piezoelectric sensor causes a current to flow through the RL circuit. The circuit is tuned to the main frequency of the vibrating structure. This causes a loss of energy in the system and thus helps in suppression of the vibrations. The design of the RL circuit is crucial for this type of vibration control, and many investigators have studied the various issues arising in the effective implementation of the shunting circuit [42–44].

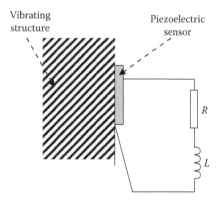

FIGURE 4.47
Schematic diagram of a passive vibration control using a piezoelectric sensor.

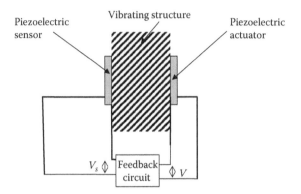

FIGURE 4.48
Schematic diagram of an active vibration control using a piezoelectric sensor and actuator.

In active damping, both a piezoelectric actuator and a sensor are used with a feedback control system. The actuator and sensor elements are bonded to the vibrating structure. The piezoelectric sensor senses the vibrations and produces an electrical voltage. This voltage is suitably modified and fed back to the piezoelectric actuator to produce counter-vibrations which effectively cancel the vibrations of the structure. A schematic diagram of an active vibration control system is shown in Figure 4.48.

The sensor piezoelectric element generates a voltage (V_s) due to the vibration of the structure. The voltage is suitably modified in terms of its phase and amplitude and fed back to the actuator piezoelectric element [42]. The actuator vibrates in such a way as to minimize the vibration of the main structure. In an actual implementation, several sensors and actuators are used in suitable positions on the vibrating structure.

4.22 Structural Health Monitoring

For the safety and reliability of big mechanical and civil structures such as turbine blades, aircraft wings, space shuttles, buildings, bridges, etc., frequent inspection of their structural condition is necessary. Detection of cracks or other damages at the right time can prevent catastrophic failure and ensure safety. Commonly used inspection techniques are x-ray, ultrasonic NDT, and thermal imaging, which are time consuming and expensive. Recent technology in this field is the Structural Health Monitoring System (SHM), which is a nonintrusive real-time monitoring system consisting of a sensor network along with integrated hardware and software. Piezoelectric sensors are embedded in the structure in critical areas, and the signals sent out by the sensors are monitored using a computer. The information in the form of patterns of electrical signals is used to identify the type and extent of damage such as strains, cracks, or fractures.

One SHM technique uses embedded piezoelectric devices as both sensors and actuators and measures the impedance of the sensors as a function of frequency by applying high-frequency AC voltage to the actuators. The resulting spectrum is used as the structure's acoustic signature, which varies if the structure is damaged [45,46].

Another technique used in SHM is generation of Lamb-waves on the structure to be inspected by embedding piezoelectric active wafers at various critical areas of the structure [47]. The piezoelectric wafers act both as lamb wave generators and detectors. The lamb waves generated propagate across the structural plane, and the piezoelectric sensors detect the waves. The output of the sensors is used as the signature of the health of the structure. From the echo signals, the extent of the damage and its exact position can be determined.

4.23 Piezoelectric Sensors and Actuators in Smart Systems and Robots

A smart system is one which can perform the functions of both sensing and actuation just as living things do. Industrial and medical robots are examples of smart systems. Piezoelectric sensors and actuators are widely used in these systems.

Piezoelectric sensors that are useful in smart systems and robots are

- Tactile sensors
- Vibration sensors
- Ultrasonic detectors

Piezoelectric actuators used in smart systems and robots are

- Piezoelectric motors
- Piezoelectric benders
- Piezoelectric vibrators
- Ultrasonic generators

Tactile sensors provide the robot with the ability to sense an object in the vicinity by touch and react in a suitable manner. Ultrasonic detectors sense ultrasound reflected from nearby objects and thus sense the proximity of the objects.

Piezoelectric motors, benders (actuators), are used for movement of robots [48–50]. For smart movement of the robots, piezoelectric ultrasonic generators and detectors are used for sensing obstacles in their path and adjusting their movements accordingly to avoid collisions. Embedded piezoelectric sensors and actuators are used to sense and control the motions. Piezoelectric actuators such as graspers and precision positioners are used along with tactile sensors to interact gently with the objects to be handled.

References

1. E. Zakar et al., 2001, Process and fabrication of a lead zirconate titanate thin film pressure sensor, *Journal of Vacuum Science and Technology A*, Vol. 19. No. 2, 345–348.

2. A. Kuoni et al., 2003, Polyimide membrane with ZnO piezoelectric thin film pressure transducers as a differential pressure liquid flow sensor, *Journal of Micromechanics and Microengineering*, Vol. 13, No. 4, S103.

3. Wilcoxon Research, MEGGITT, Industrial accelerometer design, http://www.wilcoxon.com/knowdesk/Industrial%20piezoelectric%20accelerometer%20design.pdf.

4. *Piezoelectric Accelerometers, Theory and Applications*, 2001, Metra Mess-und Frequenztechnik.

5. A. S. Chu, 2002, *Harris' Shock and Vibration Hand Book*, Fifth edition, chap. 12, ed. Cyril M. Harris, Allan G. Piersol, McGraw-Hill, New York.

6. R. D. Geddar and A. M. Madni, 1999, A Micromachined Quartz Angular rate sensor for Automation and Inertial Application, BEI Technologies Inc., Sensors.

7. P. W. Loveday and C. A. Rogers, 1998, Modification of piezoelectric vibratory gyroscope resonator parameters by feedback control, *IEEE Transactions on Ultrasonics, Ferroelectrics and Frequency Control*, Vol. 45, No. 5, 1211–1215.

8. H. Abe, T. Yoshida, and K. Turuga, 1992, Piezoelectric-ceramic cylinder vibratory gyroscope, *Japanese Journal of Applied Physics*, Vol. 31, No. 1, 3061–3063.

9. Kouichi Shuta and Hiroshi Abe, 1995, Compact vibratory gyroscope, *Japanese Journal of Applied Physics*, Vol. 34, 2601–2603.

10. A. K. Singh, 2005, Vibrating structure piezoelectric hollow cylinder gyroscope, *Indian Journal of Engineering and Material Sciences*, Vol. 12, 7–11.

11. C. Li et al., 2008, Flexible dome and bump shape piezoelectric tactile sensors using PVDF-TrFE copolymer, *Journal of Microelectromechanical Systems*, Vol. 17, No. 2, 334–341.

12. J. Dargahi et al., 1998, A micromachined piezoelectric tactile sensor for use in endoscopic graspers, Victoria, BC, Canada, Oct. 1998. *Proceedings of the International Conference on Intelligent Robots and Systems*.

13. E. S. Kolesar and C. S. Dyson, 1995, Piezoelectric tactile integrated circuit sensor, *Journal of Vacuum Science and Technology*, A13, 1001.

14. G. Murali Krishna and K. Rajanna, 2004, Tactile sensor based on piezoelectric resonance, *IEEE Sensors Journal*, Vol. 4, No. 5, 691–697.

15. E. S. Leland, E. M. Lai, and P. K. Wright, 2004, A self-powered wireless sensor for indoor environmental monitoring, Department of Mechanical Engineering, University of California, Berkeley.

16. S. Sherrit, 2008, The Physical Acoustics of Energy Harvesting, Jet Propulsion Laboratory, California Institute of Technology, IEEE International Ultrasonics Symposium Proceedings.

17. D. Charnegie, 2007, Frequency tuning concepts for piezoelectric cantilever beams and plates for energy harvesting, M. S. thesis, University of Pittsburgh.

18. M. Pozzi and M. Zhu, 2011, Plucked piezoelectric bimorphs for energy harvesting applications, *Smart Sensors, Actuators, and MEMS V*, ed. Ulrich Schmid, José Luis Sánchez-Rojas, Monika Leester-Schaedel, Proc. of SPIE, Vol. 8066.

19. B. S. Lee, S. C. Lin, and W. J. Wu, 2010, Fabrication and evaluation of a MEMS piezoelectric bimorph generator for vibration energy harvesting, *Journal of Mechanics*, Vol. 26, No. 4.

20. H. W. Kim et al., 2004, Energy harvesting using a piezoelectric "cymbal" transducer in dynamic environment, *Japanese Journal of Applied Physics*, Vol. 43, No. 9A, 6178–6183.

21. N. S. Shenck and J. A. Paradiso, 2005. Energy scavenging with shoe-mounted piezoelectrics, *IEEE Micro*, http://www.computer.org/micro/homepage/may_june/shenck/shenck_print.htm.

22. M. Loreto Mateu Saez, 2009, Energy harvesting from human passive power, Doctoral thesis, Universitat Polit_ecnica de Catalunya.

23. A. Dogan and R. Newnham, 1998, U.S. Patent 5,729,077.

24. R. J. Meyer, Jr. et al., 2002, Design and fabrication improvements to the cymbal transducer aided by finite element analysis, *Journal of Electroceramics*, 8, 163–174.

25. C.-L. Sun et al., 2006, High sensitivity cymbal-based accelerometer, *Review of Scientific Instruments*, 77, 036109.

26. J. Zhang, 2000, Miniaturized flextensional transducers and arrays, Ph.D. thesis, Penn State University.

27. E. Maione et al., 2002, Transducer design for portable ultrasound enhanced transdermal drug delivery system, *IEEE Transactions in Ultrasonics, Ferroelectrics, and Frequency Control*, Vol. 49, 1430–1436.

28. B. Snyder et al., 2006, Ferroelectric transducer arrays for transdermal insulin delivery, *Journal of Materials Science*, Vol. 41, 211–216.

29. W. G. May, 1975, Piezoelectric Electromechanical Translation Apparatus, US Patent 3902084.

30. Discovery Technology International, 2009, DTI_Tech_Comms_PiezoMotor_UPZM02-LLLP, www.dti-piezotech.com.

31. K. Uchino, 1998, Piezoelectric ultrasonic motors: Overview, *Smart Materials and Structures*, Vol. 7, 273–285; IOP Publishing Ltd.

32. T. Fabien and F. Enrico, 2006, Modelling of annular piezoelectric motors, Faculte polytechnique de Mons, Belgium. http://www.geniemeca.fpms.ac.be/Recherche/Articles/Thie2006.pdf.

33. E. Bekiroglu, 2008, Ultrasonic motors: Their models, drives, controls and applications, *Journal of Electroceramics*, Vol. 20, No. 3-4, 277–286, DOI: 10.1007/s10832-007-9193-4. From the issue entitled "Special Issue on the International Workshop of Piezolelectric Materials and Actuator Applications"; Guest Editors: Aydin Dogan, Erman Uzgur, and Kenji Uchino.

34. J. G. Smits, 1990, Piezoelectric micropump with 3 valves working peristaltically, *Sensors Actuators A*, Vol. 21, 203–206.

35. http://ndeaa.jpl.nasa.gov/nasa-nde/usdc/usdc.htm, 2002.

36. L. N. Domm, Y. Bar-Cohen, and S. Sherrit, 2011, NASA USRP—Internship Final Report, Jet Propulsion Laboratory, California Inst. of Technol.
37. Y. Bar-Cohen et al., 2001, Ultrasonic/sonic driller/corer (USDC) as a sampler for planetary exploration, *IEEE Aerospace Conference on 'Missions, Systems and Instrumentation for In Situ Sensing*, Montana.
38. X. Bao et al., 2003, *IEEE Transactions on Ultrasonics, Ferroelectrics and Frequency Control*, Vol. 50, No. 9, 1147–1160.
39. C. Y. Lin, 1997, Design and analysis of piezoelectric transformer converters, Ph.D. Dissertation, Virginia Tech.
40. W. Huang, 2003, Design of a radial mode piezoelectric transformer for a charge pump electronic ballast with high power factor and zero voltage switching, M. S. thesis, Virginia University.
41. Laoratanakul, 2007, Characteristics of Radial-mode Piezoelectric Transformers, Materials Forum, 31, Ed. J. M. Cairney and S. P. Ringer, Institute of Materials Engineering Australasia Ltd.
42. S. O. Reza Moheimani, 2003, A survey of recent innovations in vibration damping and control using shunted piezoelectric transducers, *IEEE Transactions on Control Systems Technology*, Vol. 11, No. 4, 482–494.
43. A. J. Fleming and S. O. Reza Moheimani, 2005, Control orientated synthesis of high-performance piezoelectric shunt impedances for structural vibration control, *IEEE Transactions on Control Systems Technology*, Vol. 13, No. 1.
44. S. Behrens, 2004, Vibration control using shunted piezoelectric and electromagnetic transducers, Ph.D. Thesis, The University of Newcastle, Australia.
45. H. A. Winston et al., 2001, Structural health monitoring with piezoelectric active sensors, *Journal of Engineering for Gas Turbines and Power*, Vol. 123, No. 2, 353–357.
46. G. Park et al., 2000, Impedance-based health monitoring of civil structural components, *Journal of Infrastructure Systems*, Vol. 6, No. 4, 153–160.
47. V. Giurgiutiu, 2002, Lamb Wave Generation with Piezoelectric Wafer Active Sensors for Structural Health Monitoring, 8th Annual International Symposium on NDE for Health Monitoring and Diagnostics, 2–6 March 2002, San Diego.
48. Z. Wu et al., 1992, Light-weight robot using piezoelectric motor, sensor and actuator, *Smart Materials and Structures*, Vol. 1, No. 4, 330–340.
49. M. Vogel and F. Vajda, Smart actuators for miniature mobile robots, med.ee.nd.edu/MED9/Papers/Robotics/Robotics_2/med01-069.pdf.
50. D. Campolo et al., Development of piezoelectric bending actuators with embedded piezoelectric sensors and micromechanical flapping mechanisms, http://robotics.eecs.berkeley.edu/.

5

Medical Applications of
Piezoelectric Materials

5.1 Introduction

Piezoelectric materials have a large number of medical applications based on their sensor and actuator characteristics and on their ability to generate and detect ultrasound. Many piezoelectric devices are routinely used in hospitals for medical diagnosis and therapy. The high potential of piezoelectric material as a smart material has made it a very important subject of interest for researchers all over the world, and many new innovative medical applications of the materials are constantly being reported. Table 5.1 summarizes the various piezoelectric applications/devices in the medical field. Some of the devices are commercially available, and some are still in the research and development stage.

A brief description of each of the devices/applications is given in the following sections of the chapter. In medical jargon, the term *ultrasound* is used more often than *ultrasonic*. In the following discussions, both these terms are used to mean ultrasonic field.

5.2 Blood Pressure Monitor

The principle of a piezoelectric pressure sensor and the structure of a MEMS piezoelectric pressure sensor are described in Section 4.3 of Chapter 4. The pressure sensor can be used for measuring blood pressure or for continuously monitoring blood pressure during heart surgery. The piezoelectric material used in the sensor is a thin film of PZT or polymer PVDF. For the measurement of blood pressure, the sensor is placed either on the wrist or the mid-arm. A proper backing and a strap with Velcro is provided to firmly grip the sensor around the wrist or the arm as shown in Figure 5.1a. The electronics required for the sensor such as voltage amplifier and filter may

TABLE 5.1

Medical Applications of Piezoelectric Materials

		Medical Device/Application
Applications based on sensor characteristics	Pressure sensor	• Blood pressure monitor • Pressure monitor in angioplasty balloon
	Sound sensor	• Heartbeat monitor (piezoelectric stethoscope)
	Tactile sensor	• Scalpel and grasper for minimal invasive surgery
	Vibration sensor (accelerometer)	• Measure of tremors in Parkinson's patients • Monitoring patient's activity • Pace maker control
Applications based on actuator characteristics	Piezoelectric pump and valves and other actuators	• Insulin pump • Infusion pump • Valves
Applications based on ultrasonic generation and detection	Diagnosis	• Medical imaging (sonography) • Bone density measurement
	Therapy	• Transdermal drug delivery • Localized drug delivery • Ablation of cancer cells • Cataract surgery • Bone healing and growth • Arthritic and joint inflammation treatment

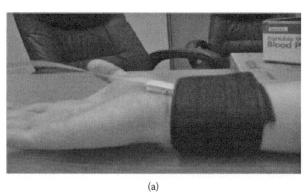 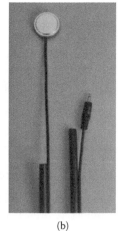

(a)   (b)

FIGURE 5.1
(a) Piezoelectric blood pressure monitor fixed on the wrist. Courtesy: Wade D. Peterson, David A. Skramsted, and Daniel E. Glumac, Reproduced from the website: www.phoenix.tc-ieee.org. (b) Pro-Wave's piezoelectric blood pressure sensor. Courtesy: Pro-Wave Electronic Corp. (website: http://www.prowave.com.tw/english/products/pp/bps.htm).

be built in or may be outside the sensor device. The analog output of the sensor may be converted to digital output display using an A/D convertor. A piezoelectric blood pressure sensor marketed by Pro-Wave is shown in Figure 5.1b.

Angioplasty, which is a surgical method of opening up blockages in the artery by inflating a balloon inside the artery, requires careful monitoring of the inflation pressure. A piezoelectric pressure sensor is used for the purpose. Other potential applications of piezoelectric pressure sensors are continuous heart rate monitoring of patients during activities such as walking, running, and treadmill exercises.

5.3 Piezoelectric Heartbeat Monitor

The piezoelectric sound sensor, which is nothing but a piezoelectric microphone, finds applications in the medical field as a heartbeat monitor. The piezoelectric microphone described in Section 4.6 of Chapter 4 can be suitably adopted for using as a heartbeat monitor. The device may have built-in amplifier or an external amplifier. The sensor in the form of a small disc is secured firmly at the chest, and the output can be observed on a CRO or may be recorded on a strip chart recorder.

5.4 Piezoelectric Tactile Sensor—Endoscopic Grasper and Minimal Invasive Surgical Instrument

Piezoelectric tactile sensors have applications in endoscopic graspers and minimal invasive surgical instruments. The principle of piezoelectric tactile sensors and their basic structure are described in Section 4.8 of Chapter 4. The tactile sensors can be made very small using micromachining techniques, and they are attached to the end of the graspers or to the tip of the scalpel. The sensor sends out electrical signals whenever it comes into contact with tissues. They are sensitive enough to detect the presence of soft tissues in the vicinity, which helps the surgeons manoeuvre the surgical instruments delicately and safely during surgery. The sensors are capable of sensing the magnitude and position of the pressure on the grasping tool or the scalpel. Tactile sensors have also applications in detection of cancerous cells through the study of distributed pressures acting on the surface.

5.5 Piezoelectric Accelerometer—Monitoring Patient Activity and Detecting Tremors

The principle and working of piezoelectric accelerometer are described in Section 4.4 of Chapter 4. As the accelerometers function as very sensitive vibration sensors, they are used to monitor activities of patients implanted with pace-makers and for detecting involuntary hand tremors in patients with neurological disorders.

A small piezoelectric accelerometer is mounted inside the pacemaker. The accelerometer gives an output proportional to the activity of the patient. The pacemaker uses the signals from the accelerometer to adjust the cardiac simulation rate. The heart rate increases when the patient activity increases and decreases when the patient is resting.

Parkinson's disease is a neurological disorder in which patients experience slight tremors in the hand and the foot on one side of the body. The frequencies of the vibrations are normally in the range of about 4–12 Hz. Piezoelectric accelerometers are useful in measuring the vibrations [1,2]. One or more accelerometers are fixed at the fingers, upper part of the hand, or at the wrist. The movements caused by the tremor result in electrical output in the accelerometers which can be recorded. Such studies help in estimating the severity of the disease and for taking decisions on the type of clinical treatment.

5.6 Piezoelectric Pump—Drug Delivery and Biomedical Analyses

The principle and working of the piezoelectric micro pump are described in Section 4.13 of Chapter 4. Micro pumps have many biomedical applications which include

- Drug delivery
- Protein analysis
- Genomics
- DNA diagnostics
- Environmental assay

The majority of micro pumps used for these applications are of the reciprocating displacement type, which use piezoelectric membranes for actuation.

Drug delivery is a promising area in which these micro pumps are very useful, and a lot of research work in this area is currently in progress. Many novel designs of piezoelectric micro pumps for drug delivery are reported in the literature [3,4]. Biocompatible micro pumps for insulin delivery which can be implanted inside the body have been tested *in vivo* on animals [5].

Most of the commercially available piezoelectric micro pumps are used mainly for biochemical analysis such as protein and DNA analysis, genomics, and environmental assays. A microfluidic chip called 'Lab-on chip' uses a micro pump for pumping fluids through micro-channel networks for bio-medical analysis.

5.7 Ultrasonic Imaging

Ultrasonic imaging is one of the most important diagnostic tools presently used in the medical field. The technique enables viewing of soft tissues of human body parts. Short pulses of high-frequency ultrasonic waves are made incident on the tissue, and the reflected pulse is detected and converted to an image that can be displayed. Piezoelectric ultrasonic generators and detectors are used for the purpose. The frequency of the ultrasonic field used is in the range of 10–50 MHz. The average intensity of the ultrasonic field is normally below 100 mW/cm^2.

The ultrasonic transducer is placed externally on the skin close to the organ which needs to be imaged. For good transmission of the ultrasonic wave through the skin to the tissue, there must be impedance matching at the transducer–skin boundary. If there is no impedance matching, more than 50% of the sound energy will get reflected at the skin. To ensure good transmission, a suitable impedance matching layer is fixed to the head of the transducer, and in addition a coupling gel is used on the skin to avoid air getting trapped between the transducer head and the skin. As the ultrasound gets transmitted, it gets attenuated due to absorption and scattering. The part of the ultrasound which reaches the organ to be imaged gets reflected, and the echo is detected by the same transducer. The transducer is scanned over the specific region to be imaged, and the detected signals from different positions are digitised and converted to grey scale. The entire image information is stored in memory and displayed on the monitor.

High-frequency ultrasonic waves give better resolution, but attenuation is higher at higher frequencies. So a compromise has to be made for imaging organs which are deep inside, such as abdomen and uterus. Frequencies in the range of 2.5–5 MHz are used for deep organs, and higher frequencies are used for superficial organs.

The approximate range of frequencies of the ultrasonic field used for imaging different organs of the body is given in Table 5.2.

TABLE 5.2

Frequency Range of Ultrasound Used for Imaging Organs of the Body

Frequency Range of Ultrasound	Organ of the Body
2.5–5 MHz	Deep abdomen and uterus
5.0–10 MHz	Breast, thyroid
10 MHz and above	Superficial veins and masses

The ultrasound is used in the form of short pulses. The shorter the pulse width, the better will be the axial resolution, and the higher the frequency, the better will be the lateral resolution. Currently used ultrasonic imaging equipment has a spatial resolution better than 1 mm.

Piezoelectric materials used for the ultrasonic transducers are ceramic PZT, polymer PVDF, or composites of PZT and polymer. Although PZT is the most sensitive piezoelectric material, it has the disadvantage of much higher impedance than that of the tissue. The polymer PVDF, although less sensitive, is preferred as it has a lower impedance to match with the impedance of the tissue. PZT–polymer composite has the advantage of having better sensitivity than the pure polymer and better impedance matching than pure PZT.

The transducer may contain a single piezoelectric element or multiple elements. Multi- element transducers are used for better focusing. A single element ultrasonic transducer used for imaging is schematically shown in Figure 5.2. The transducer consists of a piezoelectric material which is concave shaped for focusing. The piezoelectric element is supported by a backing layer which is used for damping. The backing material must have good impedance matching with the piezoelectric material for effective damping. An impedance-matching layer is attached at the head of the transducer for matching with the human tissue. The thickness of the matching layer must be one fourth of the wavelength of the ultrasonic frequency in the medium for effective transmission.

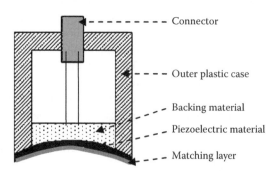

FIGURE 5.2
Schematic diagram of piezoelectric single element transducer.

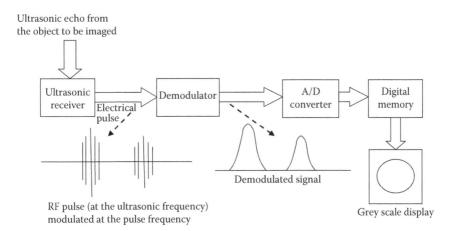

Ultrasonic echo from
the object to be imaged

RF pulse (at the ultrasonic frequency)
modulated at the pulse frequency

Demodulated signal

Grey scale display

FIGURE 5.3
Block diagram of image forming system.

The piezoelectric material (PZT or PVDF) is concave shaped for focusing the beam. The same transducer is used both as transmitter and detector. The transducer is operated at its resonance frequency, which is determined by the thickness of the piezoelectric material.

The transducer is scanned over the region to be imaged. Scanning is done automatically by rotation or oscillation. The transmitted pulse at each scan line gets reflected, and the reflected pulse is detected by the same transducer. The electrical pulse from the detector is demodulated, digitized, and stored in memory. Digital data obtained from the entire region is then reconstructed, and the grayscale image data is used for the display. The block diagram of the imaging system is shown in Figure 5.3.

5.7.1 Array Transducer

Modern ultrasonic imaging equipments use piezoelectric transducer arrays. The transducer elements are arranged in a linear array or curved array [6,7]. In the array transducer, the beam can be steered electronically by manipulating the phase of oscillation of each of the transducer elements in the array. In the linear array, the input pulse to each element is staggered or delayed successively, as illustrated in Figure 5.4. The beam can be steered by controlling the delay time between successive elements in the array.

5.7.2 Blood Flow and Tissue Motion Imaging

When ultrasound is incident on a moving object and the sound gets scattered from the object, the scattered ultrasound will undergo a shift in frequency. The shift in frequency is on the higher or lower side depending on

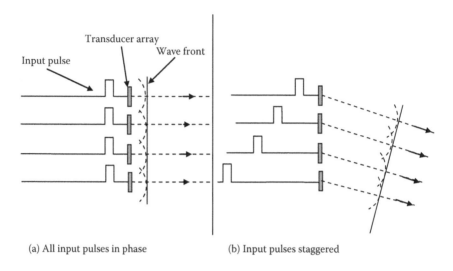

(a) All input pulses in phase   (b) Input pulses staggered

FIGURE 5.4
Linear transducer array for ultrasonic imaging. (a) Normal incidence—input pulses to the elements are in phase. (b) Steering of the beam—input pulses are staggered to deviate the beam.

whether the object is moving towards the source or away from the source, and its magnitude is proportional to the velocity of the moving object. This effect, called *Doppler shift*, is made use of in the study of blood flow and tissue motion imaging. Both continuous and pulsed ultrasounds are used in motion imaging. A piezoelectric ultrasonic transmitter and receiver placed close to each other are used for generating and receiving ultrasound.

A continuous beam of ultrasound generated from a piezoelectric transducer is passed through the body. When the beam encounters a blood vessel, it gets backscattered by the scattering particles in the blood, and the scattered ultrasound is detected by an ultrasound receiver which is kept close to the transmitter. The scattered ultrasound will have a shift in frequency because of the motion of the blood, and the shift is proportional to the component of velocity of flow in the direction of the beam. The shift in frequencies of the scattered signal from different locations of the blood vessel or a moving tissue is recorded, and the velocity distribution data are electronically processed and suitably coded to convert it to a colour image.

In pulsed ultrasound, a series of pulses are transmitted, and the echoes from the moving scatterers are detected by the receiver. The slight difference in time between the transmitted and received pulse is measured directly or the phase difference is measured from which the frequency shift can be obtained. The echo signals contain information about the motion characteristics such as speed of motion, direction of motion, and depth of the moving scatterer. These signals are coded and electronically processed to form the image of the blood vessel or the moving tissue [8–10].

5.8 Bone Density Measurement Using Ultrasound

Deterioration of bone tissue density is a common ailment affecting a large number of aged people. The disease is called osteoporosis, and it is more common in women. Early diagnosis and treatment of the disease reduces fracture risks. Two common methods used for the diagnosis of the disease are x-ray absorptiometry and quantitative ultrasound technique. The latter technique, which uses an ultrasonic transducer, is becoming more popular as it is cheaper, easy to use, and gives quite reliable diagnostic parameters.

Two principles used in the ultrasonic method are (1) measurement of attenuation of ultrasonic intensity as it transmits through the bone and (2) measurement of velocity of ultrasonic waves in the bone. In both methods, ultrasonic pulses are transmitted through the bone using an ultrasonic transmitter, and the transmitted pulse is detected using an ultrasonic receiver. The bone normally used for the measurement is the heel bone.

For both techniques, a portable ultrasonic instrument with a piezoelectric ultrasonic transmitter and receiver is used. The transmitter and the receiver are fixed on either side of the heel of the patient. For good transmission at the boundary, a gel is used between the transducer and the skin, or the foot is immersed in water. A schematic picture of the arrangement is shown in Figure 5.5. The basic structure of the ultrasonic transducers used is the same as the single-element transducer shown in Figure 5.2. The piezoelectric element need not be concave, because focusing is not required. A piezoelectric disc with a backing material and a matching layer is mounted inside a plastic case. The frequency range used normally is 200 to 1000 kHz. The thickness of the piezoelectric disc is selected to give resonance within this frequency range with a broad band output.

In the attenuation measurement method, the parameter measured is called Broadband Ultrasonic Attenuation (BUA). The attenuation of ultrasonic intensity as it enters the bone increases with increasing frequency. Ultrasonic pulses in the frequency range of 200–1000 kHz are passed through the bone.

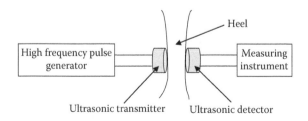

FIGURE 5.5
Bone density measurement using piezoelectric ultrasonic transmitter–detector.

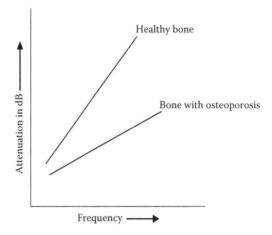

FIGURE 5.6
Attenuation vs. frequency curve for a healthy bone with osteoporosis. The slope of the line is taken as measure of bone density.

The attenuation as a function of frequency is measured through the detector. The attenuation increases linearly with frequency. The slope of the linear curve of attenuation versus frequency is called BUA and it is used as a parameter to determine the extent of osteoporosis. The slope is higher for a healthy bone than the osteoporotic bone, as illustrated in Figure 5.6. The ratio of the measured BUA to the BUA of healthy bone gives a measure of the osteoporotic condition [11].

In the velocity technique, the time elapsed between the entry of the transmitted pulse and the received pulse at the receiver is measured, and the velocity of the sound wave is measured. The velocity is higher in the healthy bone than in the osteoporotic bone. The difference in the velocity is taken as a measure of the extent of osteoporosis.

5.9 Ablation of Tumour Cells Using High-Intensity Focused Ultrasound (HIFU)

High-intensity focused ultrasound (HIFU) has been successfully used for treatment of many types of tumour. When high-intensity ultrasound is focused at a point on the human tissue, due to absorption, the tissue can attain a high temperature over a very short time. This causes damage or necrosis of the tissue at the precise area without affecting the surrounding tissues. Thus HIFU can be used for destroying tumours.

For generation of high-intensity focused ultrasound, single-element PZT transducers are not quite suitable because the PZT element needs to be concave shaped and must be quite large in size. Array transducers, which are fabricated by making deep grooves in large PZT plates, have been developed which can give high intensity and good focusing. But even these array transducers have limitations as they give rise to undesirable lateral vibrations and are susceptible to cracks due to high pressures. High-intensity transducers made of ceramic–polymer composites of 1-3 connectivity have been designed and reported by many research workers, and these transducers have several advantages over ceramic transducers [12–15]. The advantages include

- More flexible and less susceptible to mechanical damage.
- Reduced lateral vibrations because of the polymer matrix between the PZT rods.
- Thermal shaping of the structure is possible to get a concave shell.
- Higher bandwidth.

High-intensity transducers made of piezoelectric 1-3 composites in the frequency range of 200 kHz to 10 MHz with acoustic power in the range of 10–30 W/cm^2 have been developed. The number of elements in the transducer varies from about 64 to more than 200 elements. The transducers are capable of sharp focusing and generating high temperatures up to about 85°C.

The parameters of HIFU transducers such as frequency, diameter of the transducer head, the focal length, and the input power determine the amount of heat generated at the specific area. Typical HIFU transducers operate in the frequency range of 1–10 MHz and have F-numbers (ratio of focal length to head diameter) in the range of 1–1.5. The absorption of ultrasonic energy increases with increasing frequency, and so higher-frequency ultrasound generates higher temperature at the focal point. But because of high absorption, high-frequency sound waves will have lower penetration depth. Low-frequency transducers are used when deep penetration is required. Depending on the power of the transducer and the time of exposure, high temperatures, sufficient to ablate tumours, can be reached. For ablation of large tumours, the transducer must be moved continuously. The real-time image of the tumour is used for monitoring the movement of the transducer.

There are many commercially available high-intensity ultrasonic transducers. Presently, HIFU transducers are widely used for treatment of uterine fibroids and prostate cancer.

5.10 Ultrasound for Drug Delivery

In cancer treatment, after surgical removal of the primary tumour, the residual malignant cells have to be treated with systematic drug delivery. The conventional method used for drug delivery is intravenous. This method has not been very successful in delivering sufficient quantity of drug at the required site to kill all the diseased cells. Some of the reasons for ineffective drug delivery are

- The vascular endothelia of tumours are poorly developed, and so the drug may not be able to reach the required area effectively.
- The drug penetration to the tumour area through diffusion is not effective because of the high interstitial pressure of tumours.
- In brain tumours, the blood brain barrier (BBB), which protects the brain from viruses and bacteria in the blood, prevents the drug from reaching the brain.
- The drug is unable to reach all the tumour cells that are able to regenerate the tumours, and the amount of drug delivered may be insufficient to kill all the tumour cells.
- The rate at which the drug reaches the malignant cells may not be fast enough to overcome the growth rate.

Several approaches have been under investigation for enhancing the tumour reach of the anticancer drugs. Some of the approaches are drug delivery through enhanced permeability and retention, drug carrier implants, and use of nanoparticles. Especially in the treatment of brain tumours, the techniques attempted to make the drug reach the tumour site are temporary disruption of BBB, interstitial drug delivery via catheters, convection-enhanced delivery of drugs, and ultrasonic-enhanced drug delivery [16–22].

The utilization of an ultrasonic field for enhancement of drug delivery to the brain tumour is a relatively recent technique which is under investigation by many researchers. Several *in vitro* and a few *in vivo* experiments have been reported in the literature demonstrating the effectiveness of ultrasonic fields in the enhancement of drug delivery [19–22]. The mechanism believed to play a role in the enhancement is the noninertial cavitation effect, in which a large number of micro-bubbles is generated, and the bubbles oscillate (increase and decrease alternately in size) at the ultrasonic frequency. The oscillating bubbles generate mechanical stress on the walls of blood vessels without causing any harm to the tissues. This is believed to increase the vascular permeability. Other ultrasonic mechanisms believed to enhance drug delivery are acoustic streaming and sponge effect.

Piezoelectric ultrasonic transducers are used for the drug delivery. Custom-built non-focused transducers made of PZT/PVDF have been used in the experiments. Enhancement of drug intake by the brain cells in the presence of an ultrasonic field has been experimentally proved by *in vitro* experiments using brain mimic material, porcine brain, and human brain samples [19–22]. Frequencies of the ultrasound used in the experiments are in the range of 80 kHz–1.5 MHz. A pulsed ultrasonic field is used, and the time of sonification is in the range of 0.5–4 hr. The intensity of ultrasound is maintained within the threshold of noninertial cavitation (less than 1 MPa of acoustic pressure).

5.11 Ultrasonic-Induced Transdermal Drug Delivery

Transdermal drug delivery is a noninvasive technique of administering drug through the skin. The drug delivery system is in the form of patches which are adhered to the skin. The technique has advantages over oral drug delivery or injections because it makes the drug reach the body avoiding the first-pass metabolism stage and is noninvasive. Only some drugs with relatively small size molecules can be administered through the skin by using patches. Even for these drugs the method has limitations because of the low permeability of the skin.

One of the promising techniques for enhancing drug delivery through skin is the use of ultrasound [23–26]. The technique is called *sonophoresis* and is presently widely used in hospitals. Piezoelectric ultrasonic transducers are used for generating ultrasound. Some of the commercially developed sonophoresis equipments are Microlysis, Sonoderm, and SonoPrep.

The frequency of the ultrasound used for transdermal drug delivery is in the range of 20–100 KHz. Pulsed ultrasound is used as continuous ultrasound tends to heat up the area. The intensity of the ultrasound and the pulse frequency must be suitably selected depending on the type of drug and the dosage required. Single transducers or array transducers have been used.

5.12 Phaco Emulsification—Cataract Surgery

The most widely used technique for cataract surgery is phaco emulsification, in which an ultrasonic field is used to fragment the affected lens into pieces. The instrument consists of a small piezoelectric portable ultrasonic transducer with a sharp titanium micro-tip, an irrigation–aspiration system with micro-channels, a liquid container, and the required electronic control system. The hand-held transducer is inserted into the eye through a very small incision made through the cornea. The transducer is operated at a frequency of about 40 kHz. The ultrasonic field is pulsed with pulse frequencies in the range of 4–400 Hz to generate bursts of ultrasonic field at the titanium tip. The vibrating micro-tip breaks the cataract-affected lens into

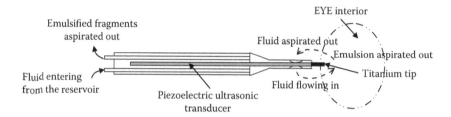

FIGURE 5.7
Schematic diagram of phaco emulsification device.

small fragments. Simultaneously, a fluid is made to enter the eye. The emulsified fragmented pieces are then aspirated out along with the cortical matter. The schematic diagram of the transducer used in phaco-emulsification is shown in Figure 5.7.

5.13 Therapeutic Ultrasound—Treatment of Injury, Muscular Pain, and Bone Fracture

An ultrasonic field of low intensity generated by a piezoelectric transducer is used for treatment of injuries and for bone healing. There have been attempts by researchers to show scientific evidence of the bone-healing properties of ultrasound [27–29]. But to this day there has not been a conclusive report to prove that ultrasound has healing effects. In most practical cases of the application of ultrasound for healing fractures or relieving muscular pains, the method is used along with other conventional treatments.

Frequencies of the ultrasound used in the treatments are in the range of 1—3MHz, and the intensity is quite low, less than 3 W/cm^2. Both continuous ultrasound and pulsed ultrasound have been in use. The time of treatment is a few minutes, the exact time depending on the type of injury or the cause of pain.

Some of the theories proposed for confirming the therapeutic effects of ultrasound are

- High-frequency vibrations passing through the tissue generate mild heat within the tissue; the heating helps to reduce pain and promote the healing process.
- Ultrasound has a positive effect on collagen by improving its extensibility; thus, it can help in the healing of injured tissues.
- Ultrasound speeds metabolism and increases blood flow.

Injuries occurring during sports require fast healing. In such cases, ultrasonic treatment is given, which is believed to help in reducing edema and simulating the healing process. Physiotherapists use ultrasonic devices to treat joint and muscular pain and for treatment of arthritis. The vibrations caused by ultrasound are believed to have a massaging effect and at the same time mildly heat up the tissues, which helps in relieving the pain.

There are many commercially available ultrasound portable devices for treatment of injuries and joint and muscular pains.

References

1. P. Henrique, G. Mansur et al., 2007, A review on techniques for tremor recording and quantification, *Critical Reviews in Biomedical Engineering*, Vol. 35, No. 5, 343–362.
2. G. Grimaldi and M. Manto, 2010, Neurological tremor: Sensors, signal processing and emerging applications, *Sensors*, Vol. 10, 1399–1422.
3. B. Ma et al., 2006, A PZT insulin pump integrated with a silicon micro needle array for transdermal drug delivery, *Microfluidics and Nanofluids*, Vol. 2, No. 5, 417–423.
4. W. J. Spencer et al., 1978, An electronically controlled piezoelectric insulin pump and valves, *IEEE Transactions on Sonics and Ultrasonics*, Vol. 25, No. 3, 153–156.
5. G. Liu et al., 2010, A disposable piezoelectric micropump with high performance for closed-loop insulin therapy system, *Sensors and Actuators A: Physical*, Vol. 163, No. 1, 291–296.
6. J. M. Cannata et al., 2006, Development of a 35-MHz piezo-composite ultrasound array for medical imaging, *IEEE Transactions on Ultrasonics, Ferroelectrics, and Frequency Control*, Vol. 53, 1, 224–236.
7. A. Nguyen-Dinh et al., 1996, High Frequency Piezo-Composite Transducer Array Designed for Ultrasound Scanning Applications, IEEE Ultrasonics Symposium-943.
8. P. N. T. Wells, 2006, Ultrasound imaging, *Physics in Medicine and Biology*, Vol. 51, R83–R98.
9. P. R. Hoskins and W. N. McDicken, 1997, Colour ultrasound imaging of blood flow and tissue motion, *The British Journal of Radiology*, Vol. 70, 878–890.
10. C. Kargel et al., 2004, Doppler ultrasound systems designed for tumor blood flow imaging, *IEEE Transactions on Instrumentation and Measurement*, Vol. 53, No. 2, 524–536.
11. M. S. Holi et al., 2005, Quantitative ultrasound technique for the assessment of osteoporosis and prediction of fracture risk, *Journal of Pure and Applied Ultrasonics*, Vol. 27, 55–60.
12. G. Fleury et al., 2002, New piezocomposite transducers for therapeutic ultrasound, 2nd International Symposium on Therapeutic Ultrasound, Seattle.
13. D. Melodelima et al., 2009, Thermal ablation by high-intensity-focused ultrasound using a toroid transducer—results of animal experiments, *Ultrasound in Medicine and Biology*, Vol. 35, 425–435.

14. G. Xuecang et al., 1999, 1-3 Piezoelectric Composites for High Power Ultrasonic Transducer Applications, IEEE Ultrasonics Symposium-1191.

15. R. Seip et al., 2004, Annular and Cylindrical Phased Array Geometries for Transrectal High-Intensity Focused Ultrasound (HIFU) Using PZT and Piezocomposite Materials, 4th International Symposium on Therapeutic Ultrasound, Kyoto, Japan, 18–20 Sept.

16. M. S. Lesniak and H. Brem, 2004, Targeted therapy for brain tumours, *Nature Reviews—Drug Discovery*, Volume 3, 499–508.

17. M. T. Krauze et al., 2008, Safety of real-time convection-enhanced delivery of liposomes to primate brain: A long-term retrospective, *Experimental Neurology*, Vol. 210, No. 2, 638–644.

18. W. A. Vandergrift and S. J. Patel, 2006, Convection-enhanced delivery of immunotoxins and radioisotopes for treatment of malignant gliomas, *Neurosurgical Focus*, Vol. 20, No. 4, E13.

19. G. K. Lewis Jr., W. L. Olbricht, and G. Lewis, 2007, Acoustic enhanced Evans blue dye perfusion in neurological tissues, 154th Meeting Acoustical Society of America, New Orleans, Louisiana. Biomedical Ultrasound/Bioresponse to Vibration.

20. V. Frenkel, 2008, Ultrasound mediated delivery of drugs and genes to solid tumours, *Advanced Drug Delivery Reviews*, Vol. 60, No. 10, 1193–1208.

21. W. G. Pitt, G. A. Husseini, and B. J. Staples, 2004, Ultrasonic drug delivery—a general review, *Expert Opinion on Drug Delivery*, Vol. 1, No. 1, 37–56.

22. Y. Liu et al., 2010, Ultrasound-enhanced drug transport and distribution in the brain, *AAPS Pharm SciTech*, Vol. 11, No. 3, 1005–1017.

23. N. B. Smith, 2007, Perspectives on transdermal ultrasound mediated drug delivery, *International Journal of Nanomedicine*, Vol. 2, No. 4, 585–594.

24. K. Saroha, B. Sharma, and B. Yadav, 2011, Sonophoresis: An advanced tool in transdermal drug delivery system, *International Journal of Current Pharmaceutical Research*, Vol. 3, No. 3, 89–97.

25. E. Maione et al., 2002, Transducer design for a portable ultrasound enhanced transdermal drug-delivery system, *IEEE Transactions on Ultrasonics, Ferroelectrics, and Frequency Control*, Vol. 49, No. 10, 1430–1436.

26. O. M. Al-Bataineh et al., 2011, Noninvasive transdermal insulin delivery using ultrasound transducers, *1st Middle East Conference on Biomedical Engineering (MECBME)*, 31-94Feb 2011, Shangah.

27. L. Claes and B. Willie, 2007, The enhancement of bone regeneration by ultrasound, *Progress in Biophysics and Molecular Biology*, Vol. 93, 384–398.

28. K. N. Malizos et al., 2006, Low-intensity pulsed ultrasound for bone healing: An overview, Injury, *International Journal of the Care of the Injured*, 37S, S56–S62.

29. V. C. Protopappas et al., 2008, Ultrasonic monitoring of bone fracture healing, *IEEE Transactions on Ultrasonics, Ferroelectrics, and Frequency Control*, Vol. 55, No. 6, 1243–1255.

6

Modelling and Virtual Prototyping of Piezoelectric Devices Using Finite Element Software Tools

6.1 Introduction

Design and fabrication of piezoelectric devices involve many experimental trials, which are expensive and time consuming. Theoretical analysis of a device using finite element methods (FEM) helps in designing a particular device for best performance. Many finite element software tools are commercially available that can be used for creating virtual prototypes and computer simulations. The use of this software helps in optimizing experimental parameters without going through expensive and time-consuming practical testing. Testing may be carried out virtually with various designs and new materials.

Piezoelectric devices require a coupled field finite element technique, as both mechanical and electric fields are involved.

This chapter gives a brief introduction to finite element techniques and the theory of coupled field systems. The procedures for coupled field analysis using software tools are discussed, and the results of FE analyses of some selected piezoelectric devices are included. In the examples given in this chapter, software tools ANSYS and PAFEC are used.

6.2 Finite Element Method

The finite element method (FEM) is a powerful tool for analyzing complex engineering systems using numerical techniques. The basic concept of FEM is that the whole structure to be analyzed may be divided into a large number of small elements of finite dimensions called *finite elements*. The whole structure is an assemblage of these finite elements, which are connected to each other by a finite number of joints called *nodes*. The governing equations of the problem are formulated for each element. The variables are expressed in terms of their values at the nodal points using a suitable polynomial. The equations formed for individual elements are then combined to form equations for the entire structure, ensuring continuity at the common nodal points. The necessary boundary conditions for the problem are applied and the equations are solved for the variables.

6.3 Theory of Coupled Field Finite Element Analysis for Piezoelectric Structure

NOTE: In this section, the notation used for strain is *s* instead of *x* as was used in earlier sections. The notation *x* is used for the X-coordinate of the Cartesian coordinate system; *s* in this section is not to be confused with the *s* used earlier for elastic compliance constant. *S* is used for the compliance constant.

Piezoelectric analysis involves both electrical and mechanical parameters. Mechanical parameters of interest are stress, strain, and pressure; electrical parameters of interest are charge, voltage, and electric field.

The governing equations of piezoelectric behaviour for direct and indirect piezoelectric effects are (Chapter 2, Equations 2.24 and 2.25b)

Direct effect:

$$D = es + \varepsilon^s E \tag{6.1}$$

Indirect effect:

$$X = c^E s - eE \tag{6.2}$$

where *s* is the *strain tensor* and the other notations are the same as before.

The electric field in terms of electric potential φ is given by

$$E = -\nabla\phi \tag{6.3}$$

The strain tensor *s* is related to the displacement *u* by

$$s = Bu \tag{6.4}$$

where

$$B = \begin{bmatrix} \partial/\partial x & 0 & 0 \\ 0 & \partial/\partial y & 0 \\ 0 & 0 & \partial/\partial z \\ \partial/\partial y & \partial/\partial x & 0 \\ 0 & \partial/\partial z & \partial/\partial y \\ \partial/\partial z & 0 & \partial/\partial x \end{bmatrix} \qquad (6.5)$$

The mechanical behaviour of piezoelectric material is described by the equation

$$divX = \rho \frac{\partial^2 u}{\partial t^2} \qquad (6.6)$$

where ρ is the density of the piezoelectric material.

The electrical behaviour of the piezoelectric material is described by the equation

$$divD = 0 \qquad (6.7)$$

because there are no free charges in the piezoelectric medium.

Equations 6.1–6.7 describe piezoelectric behaviour, and they can be solved using suitable boundary conditions for mechanical and electrical parameters.

6.3.1 Finite Element Method

In FE analysis, the piezoelectric medium is divided into a number of small discrete elements called *finite elements*. Each element has a set of interconnecting points on its edges called the *nodes*. The variables, called the *degrees of freedom* (DOF), such as displacement, potential, etc., at any arbitrary point within an element are expressed in terms of their values at the nodal points, using a suitable polynomial interpolation function (N_i) defined for each variable [1].

Figure 6.1a shows an example where a portion of the piezoelectric medium in the device is divided into number of finite elements. The type of element selected is a two-dimensional element with eight nodes as shown in Figure 6.1b.

For an element with n nodes, the displacement of a point (x, y, z), denoted by $u(x,y,z)$, is expressed in terms of the nodal displacement values $\hat{u}(x_i, y_i, z_i)$ by

$$u(x,y,z) = N_u \hat{u}(x_i, y_i, z_i) \qquad i = 1, 2, \dots, n \qquad (6.8)$$

where n is the number of nodes in the element, and N_u is the interpolation function for displacement. The interpolation function is a function of a set of natural coordinates ξ, η, ϕ defined within an element.

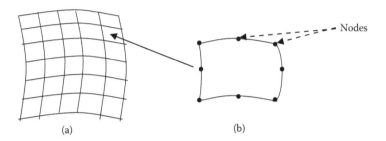

FIGURE 6.1
(a) A portion of the piezoelectric medium subdivided into finite elements (nodes not shown).
(b) Eight-noded finite element.

The potential at the point (x,y,z), denoted by $\varphi(x,y,z)$, is expressed in terms of the nodal potential values $\hat{\varphi}(x,y,z)$ by

$$\varphi(x,y,z) = N_\varphi \hat{\varphi}(x_i, y_i, z_i) \qquad i = 1, 2, \ldots, n \tag{6.9}$$

where N_φ is the interpolation function for potential.

The variational method of approximation is applied to Equations 6.6 and 6.7. Galerkin approach is used where the weight functions are equal to the interpolation functions N_i [1,2].

From Equations 6.4 and 6.8 we can write

$$s = Bu = BN_u \hat{u} = B_u \hat{u} \tag{6.10}$$

From Equations 6.3 and 6.9 we can write

$$E = -\nabla\varphi = -\nabla\left(N_\varphi \hat{\varphi}\right) = -B_\varphi \hat{\varphi} \tag{6.11}$$

Using Equations 6.6 and 6.7 the residuals of the approximations are obtained and the weighted integrals are formed:

$$\int_\Omega N_u \left[\nabla \bullet X - \rho \frac{\partial^2 u}{\partial t^2}\right] d\Omega = 0 \tag{6.12}$$

$$\int_\Omega N_\varphi \left[\nabla \bullet D\right] d\Omega = 0 \tag{6.13}$$

where N_u and N_φ are the weight functions.

Substituting for X and D from Equations 6.2 and 6.1 and using Equations 6.8, 6.9, 6.10, and 6.11 in Equations 6.12 and 6.13, we get

$$\int_\Omega N_u \left[\nabla \bullet \left(c^E B_u \hat{u} + e B_\varphi \hat{\varphi} \right) \right] d\Omega - \int_\Omega N_u \rho \frac{\partial^2}{\partial t^2} \left(N_u \hat{u} \right) d\Omega = 0 \qquad (6.14)$$

$$\int_\Omega N_\varphi \left[\nabla \bullet \left(e B_u \hat{u} - \varepsilon^S B_\varphi \hat{\varphi} \right) \right] d\Omega = 0 \qquad (6.15)$$

Integrating Equations 6.14 and 6.15 by parts and using the boundary conditions for the problem, the two equations reduce to

$$m\ddot{\hat{u}} + d_{uu}\dot{\hat{u}} + k_{uu}\hat{u} + k_{u\varphi}\hat{\varphi} = f_B + f_S + f_P \qquad (6.16)$$

$$k_{u\varphi}\hat{u} + k_{\varphi\varphi}\hat{\varphi} = q_s + q_P \qquad (6.17)$$

$\hat{u}, \dot{\hat{u}}$, and $\ddot{\hat{u}}$ are the nodal displacements, velocities, and accelerations, respectively, and $\hat{\varphi}$ are the nodal potentials. The second term in the LHS of Equation 6.16 is included to account for the mechanical damping.

m is the mass matrix:

$$m = \iiint \rho N_u{}^t N_u dV \qquad (6.18)$$

d_{uu} is the mechanical damping matrix:

$$d_{uu} = \alpha \iiint \rho N_u{}^t N_u dV + \beta \iiint B_u{}^t c^E B_u dV \qquad (6.19)$$

α and β are the damping constants.

k_{uu} is the mechanical stiffness matrix:

$$k_{uu} = \iiint B_u{}^t c^E B_u dV \qquad (6.20)$$

$k_{u\varphi}$ is the piezoelectric coupling matrix:

$$k_{u\varphi} = \iiint B_u{}^t e \, B_\varphi dV \tag{6.21}$$

$k_{\varphi\varphi}$ is the dielectric stiffness matrix:

$$k_{\varphi\varphi} = \iiint B_\varphi{}^t \varepsilon^s B_\varphi dV \tag{6.22}$$

f_B, f_S, and f_P are the external body, surface, and point forces on the element, respectively, and q_S and q_P are the electrical surface and point charges on the element, respectively. Equations 6.16 and 6.17 and all the above matrices and vectors are defined for a single element.

The equation for the whole piezoelectric body is obtained by assembling the vectors and matrices of all the single elements in the entire meshed structure.

The FE equations for the whole structure are written as

$$M\ddot{u} + D_{uu}\dot{u} + K_{uu}u + K_{u\varphi}\varphi = F_B + F_s + F_p \tag{6.23}$$

$$K_{u\varphi}u + K_{\varphi\varphi}\varphi = Q_s + Q_p \tag{6.24}$$

In Equations 6.23 and 6.24, the quantities u, ϕ, F_B, F_s, Q_s, and Q_p are the globally assembled field quantities. The matrices M, D_{uu}, $K_{u\phi}$, $K_{u\phi}$, and K_{QQ} are globally assembled matrices.

If the total number of nodes in the whole piezoelectric body is n_{total}, the matrix Equation 6.23 will consist of ($3 \times n_{total}$) linear equations because the mechanical parameters are vectors with three components each. The matrix Equation 6.24 will consist of n_{total} linear equations because the electrical parameters' potential and charge are scalar quantities with only one component each.

If the body is non-piezoelectric, the coupling matrix $K_{u\varphi}$ in Equations 6.23 and 6.24 is zero; then the two equations are independent, Equation 6.23 describing a pure mechanical system and Equation 6.24 describing a pure electrical system. The two systems are coupled through $K_{u\varphi}$, which is nonzero for a piezoelectric medium.

Solving Equations 6.23 and 6.24 yields the value of the displacement u and the electrical potential φ at various points in the piezoelectric medium. Using the u and φ values, the strain and the electric field in the medium can be computed from Equations 6.10 and 6.11.

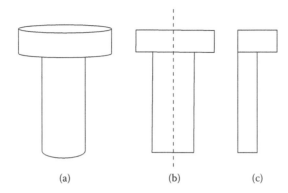

(a)   (b)   (c)

FIGURE 6.2
(a) 3-D model of a simple structure. (b) 2-D model of the structure. (c) Axisymmetric model of the structure.

6.4 Introduction to FE Analysis of Piezoelectric Devices Using a Software Tool

Analysis of a system using an FE software tool involves three processes:

  i. Preprocessing
 ii. Solving
iii. Postprocessing

6.4.1 Preprocessing

Preprocessing involves the following steps:

- Creation of a model of the structure.

  The structure may be modelled in three dimensions (3-D) or two dimensions (2-D). The two-dimensional model may be simplified if there is axial symmetry in the structure. This is illustrated in Figure 6.2 for a simple structure.

- Selection of suitable elements from the library of elements provided in the software.

  Types of finite elements commonly used are

  - 2-D three-noded triangular elements
  - 2-D six-noded curvilinear triangular elements

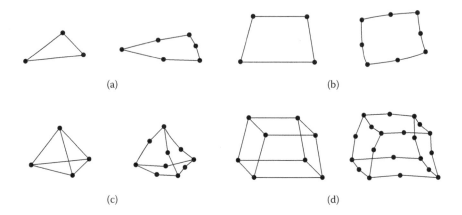

FIGURE 6.3
Standard finite elements: (a) 2-D triangular elements, (b) 2-D quadrilateral elements, (c) 3-D tetrahedral elements, and (d) 3-D rectangular prism or brick elements.

- 2-D four-noded quadrilateral elements
- 2-D eight-noded curvilinear quadrilateral elements
- 3-D four-noded tetrahedral elements
- 3-D ten-noded curvilinear tetrahedral elements
- 3-D eight-noded rectangular prism elements (brick elements)
- 3-D twenty-noded curvilinear rectangular prism elements (brick elements)

The elements are illustrated in Figure 6.3.

Each node of a piezoelectric element has 4 DOFs (viz., three displacement components u_x, u_y, and u_z and voltage φ).

- Meshing the structure with the elements.

  The number of elements used for meshing the structure determines the resolution of the solution. The optimum number of elements needs to be used. In general the element dimensions must be much less than the shortest acoustic wavelength (highest frequency) of interest. The condition to be satisfied is at least 8 to 20 elements per shortest acoustic wavelength. The aspect ratio of the elements may be close to one but, depending on the constraints of the geometry of the model, it may be even 2 or 3. Figure 6.4 shows example of a simple 2-D model of a structure meshed using eight-noded curvilinear quadrilateral elements.

- Material properties.

  Material properties of interest in the analysis of piezoelectric materials are

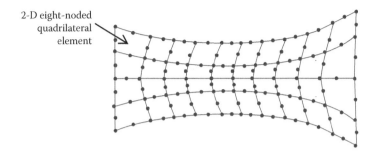

2-D eight-noded
quadrilateral
element

FIGURE 6.4
2-D model of a simple structure meshed using 2-D eight-noded quadrilateral elements.

  i. Components of piezoelectric coefficient matrix ([d] or [e])
  ii. Components of dielectric constant matrix
  iii. Components of elastic stiffness or compliance constant matrix
  iv. Density

Mostly piezoelectric materials used in applications are poled PZT ceramics or polymers. These materials are orthotropic and so the numbers of independent components of the property matrices are as given in Equations 2.40, 2.41, 2.42, and 2.43 of Section 2.4 of Chapter 2.

- *Electroding*:

  The electroded faces of the piezoelectric material must have common potential values. They are modelled by repeating the electrical DOF potential on all the nodes on the faces.

- *Boundary conditions*:

  The piezoelectric medium requires both mechanical and electrical constraints to be specified.

  i. Mechanical constraints: The mechanical DOF displacement components of all the nodes on the clamped parts of the structure are made zero.

  ii. Electrical constraints: The electrical DOF potential on all the nodes on the electroded faces of the piezoelectric material which need to be earthed are made zero.

- *External stimulus or "load"*:

  The external stimulus is either mechanical or electrical depending on the application of the piezoelectric device.

  i. Mechanical stimuli: stress, pressure, or displacement

  ii. Electrical stimuli: voltage or charge

  All the nodes on the electroded faces of the piezoelectric material on which voltage has to be applied are made to have a common voltage V.

- *Types of analysis*:

  The types of analyses of interest are
  i. Modal analysis
  ii. Harmonic analysis
  iii. Transient analysis

In modal analysis, natural frequencies of vibration and corresponding mode shapes of the system are obtained. The analysis gives information about the general dynamic behaviour of the system.

In harmonic analysis the response of the system to an external sinusoidal load (voltage or stress) at specific frequencies can be analyzed. The frequencies at which the calculations must be made are specified in the program. There are provisions to include suitable damping conditions.

In transient analysis the response of the system to an external load which has more general time dependence can be analyzed. For example, the external load (stress or voltage) may be a pulse of specific time duration and shape.

6.4.2 Solving

Solving involves selection of the type of analysis and solving the governing equations. The software provides facilities for solving the equations for a set of given boundary conditions and specified external stimulus.

6.4.3 Postprocessing

Postprocessing involves

- Acquiring results of modal analysis, for example, natural frequencies of the structure and the corresponding mode shapes
- Obtaining response to external stimulus (e.g., voltage or stress) in terms of displacement, velocity, voltage, charge, etc. at required points on the structure
- Viewing simulation of the behaviour of the structure under different conditions
- Animations
- Plotting graphs

6.5 Examples of FE Analyses of Piezoelectric Devices

Analyses of the following piezoelectric devices using the FE software tool PAFEC/ANSYS are described in this section:

- Piezoelectric bimorph
- Piezoelectric drill
- Piezoelectric gyroscope
- Piezoelectric cymbal transducer
- Piezoelectric piston in water

6.5.1 Piezoelectric Bimorph

The principle and structure of a piezoelectric bimorph are described in Section 4.11 of Chapter 4. FE analysis of a piezoelectric bimorph using ANSYS has been reported by Fiona Lowrie et al. [3]. In this section FE analysis of a bimorph using software tool PAFEC 8.8 is described.

The bimorph connected in parallel configuration is shown in Figure 6.5.

6.5.1.1 Preprocessing

6.5.1.1.1 Modelling

3-D model of the bimorph is created.
The dimensions of the bimorph:

Length (along Z-axis) = 20 mm
Width (along Y-axis) = 3 mm
Thickness of each strip (along X-axis) = 2 mm

The clamped extent of the bimorph from one end is taken as 4 mm.

Electroding the two faces of the strips is achieved by making all the nodes on the faces to have common values of the DOF electric potential. All the nodes on the grounded face (the common face) are made to have zero potential, and all the nodes on the other two faces are made to have the applied potential *V.*

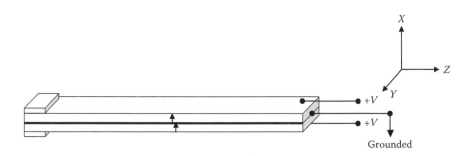

FIGURE 6.5
3-D structure of a piezoelectric bimorph connected in parallel configuration.

6.5.1.1.2 Meshing

The element used to mesh the structure is the 3-D 20-noded curvilinear brick elements shown in Figure 6.3d. The bimorph is meshed with 20 elements along its length.

6.5.1.1.3 Material Properties

The piezoelectric material selected for the bimorph is PZT 5A. Properties of the material (taken from Reference 4) are as given next.

Compliance matrix [s]:

$$
\begin{bmatrix}
16.4 & -5.74 & -7.22 & 0 & 0 & 0 \\
-5.74 & 16.4 & -7.22 & 0 & 0 & 0 \\
-7.22 & -7.22 & 18.8 & 0 & 0 & 0 \\
0 & 0 & 0 & 47.5 & 0 & 0 \\
0 & 0 & 0 & 0 & 47.5 & 0 \\
0 & 0 & 0 & 0 & 0 & 44.3
\end{bmatrix} \times 10^{-12}\,\mathrm{m}^2/N
$$

Piezoelectric coefficient [d]:

$$
\begin{bmatrix}
0 & 0 & 0 & 0 & 584 & 0 \\
0 & 0 & 0 & 584 & 0 & 0 \\
-171 & -171 & 374 & 0 & 0 & 0
\end{bmatrix} \times 10^{-12}\ C/N
$$

Permittivity matrix [ε]:

Density:

$$
\begin{bmatrix}
15.310 & 0 & 0 \\
0 & 15.310 & 0 \\
0 & 0 & 15.045
\end{bmatrix} \times 10^{-9}\ F/m
$$

$$\rho = 7750\ \mathrm{kg/m}^3$$

Normally, the piezoelectric coefficient d_{33} refers to the coefficient in the direction of polarization, which is the z-direction. But in this model the polarization direction is taken as x-direction. While entering the d-component values this should be taken care of.

6.5.1.1.4 Boundary Conditions

Mechanical: At the clamped end of the bimorph (up to a length of 4 mm) the displacements of all the nodes in x-, y-, and z-directions are constrained; that is, for all the nodes in this region DOF u_x, u_y, and u_z are made zero.

Electrical: Zero potential is applied to all the nodes on the common electroded face of the bimorph which is grounded; that is, for all the nodes on the common face DOF φ is made zero.

6.5.1.1.5 External Stimulation

A voltage of 750 V is applied to top and bottom electrodes of the bimorph; that is, for all the nodes on these two faces DOF φ is given the value 750 V.

6.5.1.2 Solving

Initially, modal analysis is carried out to get the natural frequencies of the bimorph system.

Harmonic analysis is carried out by applying a voltage of 750 V to the two faces at the first modal frequency.

6.5.1.3 Postprocessing

The first five modes of the natural frequency of the bimorph are tabulated in Table 6.1.

The first vibration mode at 3.61 kHz is the bending mode, which can be used for actuation. This mode of vibration is shown in Figure 6.6.

The displacements at the unclamped end of the bimorph, when a voltage of 750 V is applied on the outer faces at the first modal frequency (3.61 kHz), are tabulated in Table 6.2.

The analysis shows that exciting the bimorph by applying a voltage of 750 V at its natural frequency of 3.61 kHz makes the bimorph bend at its free end by about 1 mm in the x-direction. By changing the dimensions of the bimorph and the clamping conditions, higher displacements can be achieved at lower voltages.

6.5.2 Piezoelectric Drill

The piezoelectric drill is described in Section 4.13 of Chapter 4.

In Reference 5, FE analysis of the drill using ANSYS has been reported. In this section, FE analysis of the drill using the PAFEC 8.8 is described.

TABLE 6.1

The First Five Modes of Natural
Vibrations of the Bimorph

Mode No.	Frequency in kHz
1	3.61
2	5.23
3	22.74
4	23.90
5	28.89

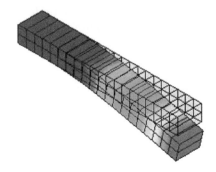

FIGURE 6.6
First natural frequency mode, which is the bending mode at 3.61 kHz.

TABLE 6.2

Displacements of the Unclamped End of the Bimorph
at the First Modal Frequency

Clamped Distance	Frequency in kHz		Displacement at Unclamped End in m
4 mm	3.61	u_x	0.9822×10^{-3}
		u_y	-0.2436×10^{-6}
		u_z	0.8291×10^{-4}

6.5.2.1 Preprocessing

6.5.2.1.1 Modelling

The picture of the piezoelectric drill is reproduced from Reference 6 in
Figure 6.7.

The drill contains the following parts:

1. Stress bolt (steel)
2. Back mass (steel)
3. Stack of piezoelectric discs
4. Horn (titanium alloy)
5. Spring (modelled as neoprene O-ring)

Four piezoelectric discs (with a central hole) are used in the stack. The discs
are stacked in such a way that alternate discs are poled in opposite directions. Each of the discs is electroded on both the faces.

The properties of the materials used in the drill are as given next (taken
from Reference 4):

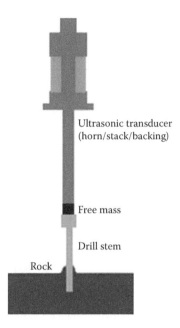

FIGURE 6.7
Ultrasonic drill designed for rock drilling. (Courtesy: X. Bao et al., 2003, *IEEE, Transactions on Ultrasonics, Ferroelectrics and Frequency Control (UFFC)*, Vol. 50, No. 9, 1147–1160.)

Compliance matrix [s]:

$$
\begin{bmatrix}
11.5 & -3.7 & -4.8 & 0 & 0 & 0 \\
-3.7 & 11.5 & -4.8 & 0 & 0 & 0 \\
-4.8 & -4.8 & 13.5 & 0 & 0 & 0 \\
0 & 0 & 0 & 31.9 & 0 & 0 \\
0 & 0 & 0 & 0 & 31.9 & 0 \\
0 & 0 & 0 & 0 & 0 & 30.4
\end{bmatrix} \times 10^{-12} \ \mathrm{m^2/N}
$$

Piezoelectric coefficient matrix [d]:

$$
\begin{bmatrix}
0 & 0 & 0 & 0 & 330 & 0 \\
0 & 0 & 0 & 330 & 0 & 0 \\
-97 & -97 & 225 & 0 & 0 & 0
\end{bmatrix} \times 10^{-12} \ \mathrm{C/N}
$$

TABLE 6.3

Properties of the Materials Used in the Drill

Material	Young's Modulus (Pa)	Density (kg/m³)	Poisson's Ratio
Steel	207×10^9	7650	0.292
Ti-alloy	114×10^9	4430	0.33
Neoprene	0.3×10^6	1150	0.46

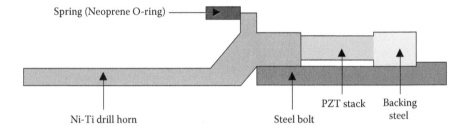

FIGURE 6.8

Axi-symmetric model of the piezoelectric drill created using PAFEC.

Permittivity matrix [ε]:

$$\begin{bmatrix} 11.416 & 0 & 0 \\ 0 & 11.416 & 0 \\ 0 & 0 & 8.85 \end{bmatrix} \times 10^{-9} \ F/m$$

Density:　　　　　　　　　$\rho = 7600 \ kg/m^3$

Properties of other materials in the drill are tabulated in Table 6.3.

The axi-symmetric model of the piezoelectric drill created using the software tool PAFEC 8.8 is shown in Figure 6.8. The dimensions of the various parts of the drill taken from Reference 5 are tabulated in Table 6.4.

6.5.2.1.2 Meshing

Eight noded curvilinear quadratic elements are used for meshing the piezoelectric rings, the bolt, and the head mass; eight-noded quadratic elements and six-noded triangular elements are used for meshing the horn of the axisymmetric model of the drill. The mesh sizes are chosen so as to have about 25–30 elements per wavelength at the highest frequency of interest (25 kHz). The meshed model of the drill is shown in Figure 6.9.

TABLE 6.4

Dimensions of the Various Parts of the Drill

Drill Part	Diameter in m	Length/Thickness in m
Stress bolt	Screw: 0.00952	Screw: 0.0339
	Head: 0.011	Head: 0.00902
Back mass	0.0268	0.0127
Horn	Base: 0.0268/0.0366	0.0683
	Tip: 0.0089	
PZT stack (4 rings)	OD: 0.025	Stack length: 0.0212
	ID: 0.0125	Thickness of each ring: 0.0053

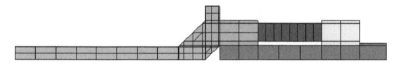

FIGURE 6.9
Axi-symmetric model of the meshed piezoelectric drill.

6.5.2.1.3 Boundary Conditions and Applied Loads

For modal analysis, all the electrodes of the PZT rings are grounded. For harmonic analysis the outermost electrodes and the middle electrodes of the PZT stack are earthed, and an alternating voltage of 1000 V is applied to the other alternate electrodes (Figure 6.10).

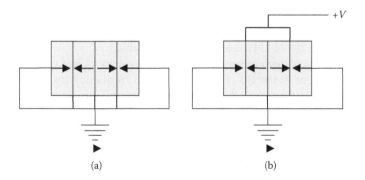

FIGURE 6.10
PZT stack with four PZT rings stacked such that alternate rings are poled in opposite directions. (a) Modal analysis: all the electrodes of the four rings grounded. (b) Harmonic analysis: the outer electrodes and the middle electrode are grounded, the other two alternate electrodes are connected, and a positive voltage is applied to the electrodes.

TABLE 6.5

The First Five Modes of the Natural
Vibrations of the Piezoelectric Drill

Mode No.	Frequency in kHz
1	12.56
2	21.44
3	30.53
4	67.02
5	73.18

6.5.2.2 Solving

In modal analysis the drill is modelled without the neoprene ring and the subspace iteration method is used [7].

Three cases are considered:

- No damping
- Damping with $\beta = 0.001$
- Damping with $\beta = 0.002$

6.5.2.3 Postprocessing

First five modes of natural vibrations of the drill are shown in Table 6.5. By studying the mode shapes, it is found that Mode 2 is the best mode for the drilling function. The resultant displacements in Mode 2 are shown in Figure 6.11.

The displacements of the tip of the horn in the axial direction, when an ac voltage of 1000 V is applied to the drill for three damping conditions, in the frequency range 20–24 kHz, are tabulated in Table 6.6.

The frequency of maximum response is close to 21 kHz. The graph of axial displacement of the tip of the drill versus frequency for zero damping ($\beta = 0$) is shown in Figure 6.12. The frequency for maximum response shifts slightly to a higher value with increasing damping. The exact frequency can be found by studying the response in a closer range near 21 kHz.

6.5.3 Piezoelectric Gyroscope

The principle and working of a piezoelectric gyroscope are described in Section 4.5 of Chapter 4.

For the finite element analysis of gyroscopes, the Coriolis effect is incorporated into the general FE Equation 6.23 by adding two more terms to the equation:

$$M\ddot{u} + (D_{uu} + G)\dot{u} + (K_{uu} - K_c)u + K_{u\varphi}\varphi = F_B + F_s + F_p \qquad (6.25)$$

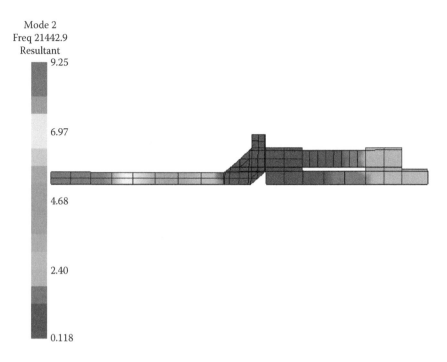

FIGURE 6.11
Modal analysis result of piezoelectric drill: Resultant displacements in mode 2 (21.44 kHz).

TABLE 6.6

Axial Displacement of the Tip of the Piezoelectric Drill

	Axial Displacement of the Horn Tip in µm		
Frequency in kHz	$\beta = 0$	$\beta = 0.001$	$\beta = 0.002$
20	5.890	5.889	5.888
21	434.8	309.2	165.7
22	−7.039	−7.038	−7.036
23	−3.824	−3.824	−3.824
24	−2.820	−2.820	−2.820

The two terms added are $G\dot{u} - K_c u$.

[G] is the damping matrix due to the Coriolis effect and $[K_c]$ is the "spin softening" effect due to the centrifugal effect.

The gyroscope selected for FE analysis is the piezoelectric vibratory cylindrical gyroscope described in Section 4.5. The gyroscope consists of a thin-walled steel cylinder to which four piezoelectric elements are symmetrically attached for actuation and sensing [8]. The FE analysis is carried out using the software tool ANSYS 10.0.

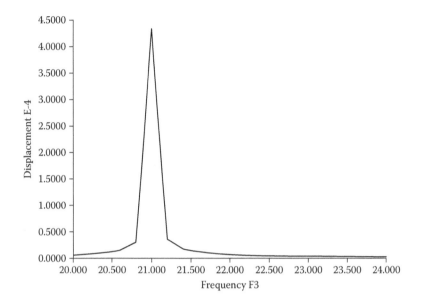

FIGURE 6.12
Frequency response of the axial displacement of the tip for β = 0.

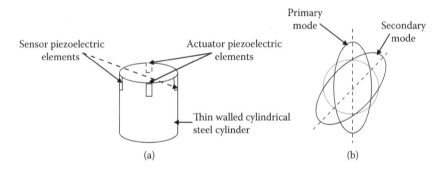

FIGURE 6.13
(a) Schematic of piezoelectric cylindrical vibratory gyroscope, and (b) the primary and secondary modes of vibration of the gyroscope.

The schematic diagram of the gyroscope is shown in Figure 6.13a.

The two modes of the vibration of the cylinder used in the vibratory gyroscope are shown in Figure 6.13b. These two modes are called the *primary* and *secondary modes*. The primary mode has antinodes at 0°, 90°, 180°, and 270°; the secondary mode has the same form as the primary mode but is rotated by 45° with respect to the primary mode. The cylinder is excited to vibrate in the secondary mode by applying sinusoidal voltage to the two piezoelectric actuator elements. This generates a voltage at the sensor elements. When the cylinder rotates about its axis, the Coriolis force couples the energies of the primary and secondary modes, which results in a change of voltage appearing at the piezoelectric sensors. The change in voltage at the sensors is proportional to the rate of rotation. The sensor voltage can be calibrated to measure the rotation rate of the cylinder. The differential output of the two sensors is taken for measurement of rotation.

6.5.3.1 Preprocessing

6.5.3.1.1 Modelling

The 3-D FE model of the gyroscope is shown in Figure 6.14. The dimensions of the cylinder are height, 12 mm; inner diameter, 12 mm; and outer diameter, 14 mm. The four piezoelectric elements of dimensions 3 × 2 × 0.25 mm are placed at the top edge of the cylinder at positions 0°, 90°, 180°, and 270°.

6.5.3.1.2 Material Properties

Properties of steel used for the cylinder:
    Two types of steel are selected for the analysis. The properties of the steel types are given in Table 6.7.

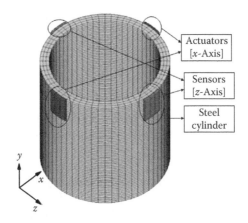

FIGURE 6.14
Finite element model of the piezoelectric cylindrical vibratory gyroscope.

TABLE 6.7

Properties of the Two Types of Steel Used for the Gyroscope

Property	Steel Type 1	Steel Type 2
Density (kg/m³)	7800	7800
Young's modulus (Pa)	190×10^9	209×10^9
Poisson's ratio	0.3	0.3

Properties of piezoelectric material:

Three types of piezoelectric materials are used for the analysis: PZT5A, PZT5H, and PZT5J. The properties of the PZT materials are taken from Reference 4.

The two faces of the four piezoelectric elements are electroded (coupled DOFs). The poling directions for the actuators and the sensors are taken normal to their faces, that is, actuator piezoelectric elements are poled along the X-direction and sensor elements are poled along the Z-direction. The poling direction is taken care in the analysis through the property matrix entry.

6.5.3.1.3 Meshing

A convergence study was carried out to fix the number of elements for the analysis. An FE model with 12,928 elements was found to be suitable.

6.5.3.1.4 Boundary Conditions and Applied Load

Mechanical constraint: Part of the bottom portion of the cylinder is constrained in all directions.

Electrical constraint: The faces of the four strips touching the cylinder are grounded ($V = 0$).

Angular velocity is applied to the cylinder about the axis of the cylinder (y-axis). A voltage of 10 V is applied to the outer faces of the two actuator piezoelectric elements.

6.5.3.2 Solving

6.5.3.2.1 Modal Analysis

Modal analysis is carried out for the gyroscope to obtain natural frequencies and to know the primary and secondary mode shapes of the structure.

6.5.3.2.2 Harmonic Analysis

Harmonic analysis is carried out by applying a sinusoidal voltage of amplitude 10 V at the secondary modal frequency and simultaneously applying angular velocity about the axis of the cylinder. The range of angular velocities applied is 0 to 200 rpm (0 to 1200 degrees/s).

6.5.3.3 Postprocessing

6.5.3.3.1 Modal Analysis

The primary and the secondary modal frequencies of the gyroscope for the different PZT materials and for the two steel types are tabulated in Table 6.8.

6.5.3.3.2 Harmonic Analysis

The gyroscope is actuated by applying sinusoidal voltage of 10 V to the two actuator piezoelectric elements.

TABLE 6.8

Primary and Secondary Modal Frequencies of the Gyroscope
for Different Steel Types and PZT Types

Steel Type	PZT Type	Primary Modal Frequency (Hz)	Secondary Modal Frequency (Hz)
Type 1	PZT 5A	17683	17757
	PZT 5H	17701	17761
	PZT 5J	17696	17762
Type 2	PZT 5A	18526	18620
	PZT 5H	18545	18625
	PZT 5J	18540	18626

Initially harmonic analysis is carried out by fixing the rotation rate of the gyroscope at 50 rpm and measuring the output voltage of the sensor element (one of the sensors) for different actuation frequencies in a frequency range close to the secondary modal frequency. The results of the analysis are shown in Figure 6.15a,b for the gyroscopes of steel types 1 and 2. From the figure it is seen that the output of the sensor is maximum for an actuation frequency of 17.7 kHz for the gyroscope of steel type 1 and 18.6 kHz for the gyroscope of steel type 2. Among the piezoelectric types used PZT 5H gives highest voltage output.

The gyroscope is actuated at the peak frequency (maximum output voltage) and the differential voltages of the two sensors are obtained for various rotation rates in the range 0–200 rpm for both clockwise and anticlockwise rotations. The results of the analysis are shown in Figure 6.16 for the gyroscope of steel type 2 and piezoelectric material PZT 5H.

Over short rotation rate ranges the graph is almost linear. The rotation rates that are useful for automobile applications are low values in the range 5–15 rpm. In this range the variation is linear.

6.5.4 Finite Element Analysis of Structures in Water

For analysis of vibrating structures in water, acoustic finite elements are required to mesh the water body surrounding the structure. The acoustic finite elements are coupled with the structural elements at the structure–water interface. The types of acoustic finite elements and the procedure for meshing the water body using them are described in the following section [9].

6.5.4.1 Acoustic FE Elements

Finite element analysis of a structure in water requires fluid finite elements for modelling the water body surrounding the structure. Fluid finite elements are called *acoustic finite elements*.

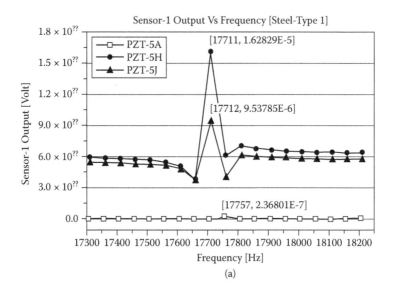

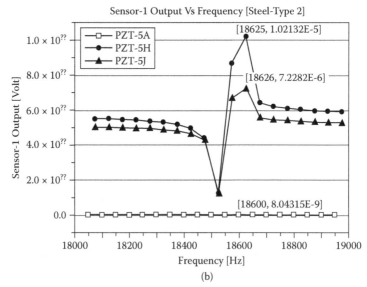

FIGURE 6.15
Output voltage of the piezoelectric sensor as a function of actuation frequencies for a fixed rotation rate of 50 rpm. (a) Steel type 1: actuation frequency range 17300–18200 Hz, and (b) steel type 2: actuation frequency range 18000–19000 Hz.

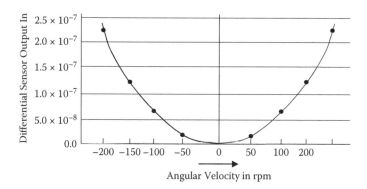

FIGURE 6.16
The differential sensor output as a function of rotation rate for gyroscope of steel type 2 and piezoelectric type PZT5H.

The fluid body is meshed using acoustic finite elements that are coupled with the structural elements. On the interface of structure and water there is continuity of normal velocity; the pressure in the fluid has continuity with the negative of normal stress on the structure.

The surface of unrestrained nodes on the outer boundary of the acoustic region is modelled as a hard surface with the boundary conditions as

normal pressure gradient $\partial p/\partial z = 0$ and velocity $\upsilon = 0$.

On the fluid structure interface, continuity of normal velocity is enforced by the equation

$$\frac{\partial p}{\partial t} = -\rho \frac{\partial^2 u_n}{\partial t^2}$$

where ρ is the density of the fluid.

FE analyses that can be carried out on a coupled structure-fluid system are

- Natural frequencies and mode shapes
- Sinusoidal response
- Transient response

6.5.4.1.1 Acoustic Elements

The types of acoustic elements are

- Pressure-based acoustic finite elements
- Acoustic boundary elements
- Displacement-based acoustic finite elements
- Wave envelope elements

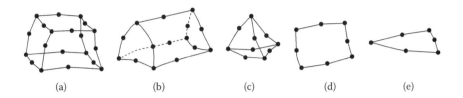

FIGURE 6.17
Acoustic finite elements.

6.5.4.1.1.1 Pressure-Based Acoustic Elements   In pressure-based acoustic elements, each node has one degree of freedom, the pressure p. Some of the commonly used fluid elements are

- 20-noded acoustic brick elements—Figure 6.17a
- 15-noded acoustic triangular prisms—Figure 6.17b
- 10-noded acoustic tetrahedral element—Figure 6.17c
- 8-noded axisymmetric quadrilateral element—Figure 6.17d
- 6-noded axisymmetric triangular element—Figure 6.17e

6.5.4.1.1.2 Acoustic Boundary Elements   Acoustic boundary elements are used to model the outer boundary of the water body surrounding the structure. They are coupled with the pressure-based fluid elements on a face-to-face basis with coincident nodes. The nodes on the interface are distinct but coincident. Each node has pressure as the DOF. The types of boundary elements commonly used are:

- Eight-noded quadrilateral element—Figure 6.18a
- Six-noded triangular patch element—Figure 6.18b
- Axi-symmetric n-noded boundary line element—Figure 6.18c

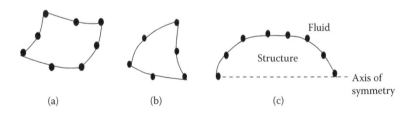

FIGURE 6.18
Acoustic boundary elements.

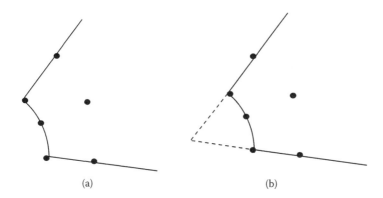

(a)   (b)

FIGURE 6.19
(a) Wave envelope element, and (b) the source point of the sides of the element.

For each boundary element, the surface defined by the patches, together with the reflections of these in the planes of symmetry defined in the model, must form a complete closed surface.

6.5.4.1.1.3 Displacement-Based Acoustic Finite Elements In displacement-based acoustic elements each node has three degrees of freedom: u_x, u_y, and u_z. Two types of diplacement-based acoustic elements are 20-noded brick elements and 15-noded triangular prism elements..

6.5.4.1.1.4 Wave Envelope Elements Wave envelope acoustic elements are used to model a fluid region in axisymmetric models of the structure–fluid systems. The elements are used to model a semi-infinite rectangle. Each element has six nodes as shown in Figure 6.19a with pressure as DOF. The sides of the elements are along the radial direction from a source point as shown in Figure 6.19b and the pressure variation along the radius is assumed to be quadratic and it extends to infinity.

The wave envelope elements are normally used along with the acoustic finite elements. The water body around the structure is first meshed with acoustic finite elements up to a certain distance ending with a surface of a sphere and from the surface onward, wave envelope elements are used to mesh the region exterior of the sphere.

6.5.4.2 Vibrating Rigid Circular Piston in Water

Acoustic radiation from a vibrating rigid circular piston is of interest in many applications, which include loud speakers, open-ended organ pipes, sonar projectors, ultrasonic medical diagnosis equipment, and physiotherapy treatment heads. In sonar projectors, the piston vibrates under water, generating ultrasonic waves, which are used for ranging and detection. The study of far field pressure produced by the vibrating piston and its radiation impedance helps in the design of acoustic transducers for sonar applications.

The pressure along the axis of a vibrating circular piston of radius "a" is given by the relation [10]

$$P_{axial}(r) = \frac{1}{2}\rho_0 c \upsilon_0 \left(\frac{a}{r}\right) ka \qquad (6.26)$$

where ρ_0 is the density of the medium, c is the velocity of acoustic waves in the medium, υ_0 is the velocity amplitude of the piston, and k is the wave vector of the acoustic wave. The above relation is valid under the condition that the distance $r \gg \pi a^2 / \lambda$.

The radiation impedance of a vibrating piston in water is of special interest for underwater acoustics applications. The real and imaginary parts of the impedance at low frequencies ($ka \ll 1$) are given by [10]

$$R \approx \frac{1}{2}\rho_0 c S (ka)^2$$

$$X \approx \left(\frac{8}{3\pi}\right)\rho_0 c S k a \qquad (6.27)$$

In this section, finite element analysis of a circular piston vibrating in water using the software tool PAFEC 8.8 is described.

6.5.4.2.1 Preprocessing

6.5.4.2.1.1 Modelling   The circular piston of radius 0.1 m is modelled using structural shell elements. Only 30° segment of the circular piston is modelled. The water medium surrounding the piston is modelled using acoustic finite elements and acoustic boundary elements.

The frequency range of the analysis is 1 kHz to 30 kHz. The 30° segment of the piston is divided into eight elements, which amounts to four elements per wavelength at the highest frequency. The thickness of the acoustic finite element is taken to be 0.025 m. The acoustic element used is a 20-noded pressure-based brick element. The acoustic boundary patches on top of the acoustic FE elements are created on a face-to-face basis. For the boundary elements, six-noded triangular and eight-noded quadrilateral patches are used. The orientation of the boundary patches are such that the positive normal points out of the fluid into the structure. The FE model of the piston in water is shown in Figure 6.20a,b.

6.5.4.2.1.2 Material Properties   PZT material used for the analysis is PZT-5A whose properties are given in Section 6.5.1. The properties of water needed

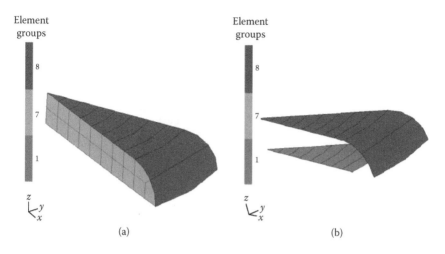

FIGURE 6.20

(a) The FE model of the piston in water (30° segment). Dark grey: piston, modelled using shell elements (not seen in the figure). Light grey: Acoustic body modelled using acoustic brick elements. Black: acoustic boundary modelled using acoustic boundary elements. (b) FE model of the piston and the acoustic boundary (acoustic body not shown).

for the analysis are density ρ and acoustic velocity in water c. The software calculates the bulk modulus B using the relation $B = \rho c^2$.

6.5.4.2.1.3 Boundary Conditions and Loading   The piston is constrained in all degrees of freedom except u_z; that is, it is permitted to vibrate only in the normal direction. The boundary conditions for the acoustic boundary element is selected to have a rigid boundary (i.e., $(\partial p / \partial n) = 0$ in the planes $y = 0$ and $z = 0$).

A uniform displacement of 2×10^{-9} m in the z-direction is specified for the circular piston. The solution type used for the FE analysis is the CHIEF formulation method [9].

6.5.4.2.2 Solving

The following analyses are carried out on the vibrating piston:

- Pressure at a distance of 1 m from the piston on the axis as a function of frequency.
- Pressure amplitude along the axis of the piston at a fixed frequency of 25 kHz

6.5.4.2.3 Postprocessing

The pressure amplitudes at a distance of one meter from the piston along the axis as a function of frequency obtained using PAFEC are tabulated in Table 6.9. The values are compared with analytical values evaluated using equation (6.26).

TABLE 6.9

Pressure Amplitude at a Distance of 1 m
on the Axis of the Piston

| | Pressure Amplitude at 1 m Distance in N/m^2 | |
Frequency (Hz)	PAFEC	Analytical (Equation 6.26)
1000	0.394	0.394
5000	9.842	9.841
10000	39.323	39.312
15000	88.276	88.251
20000	156.047	156.392
25000	242.437	243.360
30000	346.117	348.680
35000	464.128	471.774
40000	602.418	611.962

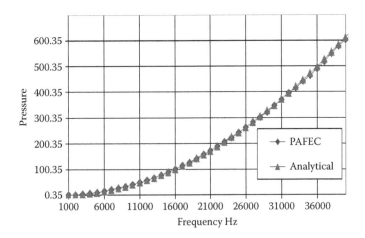

FIGURE 6.21
Axial pressure amplitude at a distance of 1 m from the piston as a function of frequency; the
values evaluated using PFEC are compared with analytical values.

The graphs of axial pressure values as a function of frequency evaluated
using FE analysis and those evaluated using the analytical equation are
shown in Figure 6.21.

The pressure amplitude along the axis of the piston as a function of distance from the piston at a fixed frequency of 25 kHz obtained using PAFEC
are tabulated in Table 6.10. The values are compared with analytical values
calculated using Equation 6.27.

TABLE 6.10

Pressure Amplitude along the Axis of the
Piston for Acoustic Wave of Frequency 25 kHz

| r/a | Normalized Pressure at 25 kHz | |
	PAFEC	Analytical
0.5	0.093	0.094
1	0.824	0.826
1.5	0.996	1.000
2	0.941	0.944
2.5	0.842	0.846
3	0.748	0.751
3.5	0.667	0.669
4	0.598	0.601
4.5	0.541	0.544
5	0.494	0.496
5.5	0.453	0.455
6	0.418	0.420
6.5	0.388	0.390
7	0.362	0.364
7.5	0.339	0.341
8	0.319	0.320
8.5	0.301	0.302
9	0.285	0.286
9.5	0.270	0.271
10	0.257	0.258

The graphs of axial pressure amplitude as a function of distance from the piston evaluated using FE analysis and those evaluated using the analytical equation are shown in Figure 6.22.

The FE analysis values obtained using PAFEC are found to agree well with the theoretical values of acoustic pressures up to a maximum frequency of 30 kHz. For frequencies higher than 30 kHz the FE mesh density must be increased to get good agreement with theory

6.5.4.3 Cymbal Transducer

The cymbal transducer described in Section 4.11 of Chapter 4 is a piezoelectric device that can be used both as an actuator and a sensor. In underwater acoustics, the cymbal transducer is used as an ultrasonic generator (projector) and a receiver (hydrophone).

In this section, an FE analysis of a double disc cymbal transducer using PAFEC software tool is described. The analysis of the transducer as an ultrasonic generator is carried out both in air and in water.

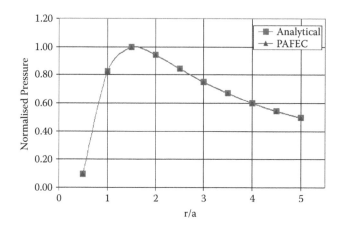

FIGURE 6.22
Pressure amplitude as a function of distance from the piston along the axis of the piston.

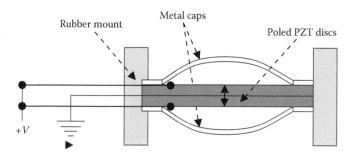

FIGURE 6.23
Double disc cymbal transducer with metal caps and rubber mount.

The cymbal transducer consists of two PZT discs poled in opposite directions held together and two cymbal shaped metal end-caps fixed on either side. The assembly is fixed in a cylindrical rubber mount. The common face of the two discs is grounded and leads are taken from the end electrodes. The schematic and 3-D diagrams of the cymbal transducer are shown in Figures 6.23 and 6.24.

6.5.4.3.1 Preprocessing

6.5.4.3.1.1 Modelling   The 2-D model of the cymbal transducer is simplified using the symmetry of the structure. The structural symmetry plane and the axi-symmetric plane of the transducer are shown in Figure 6.25a. The axi-symmetric FE model of one half of the transducer is shown in Figure 6.25b.

Meshing is done such that there are more than 20 elements per wavelength at the highest frequency of interest.

FIGURE 6.24
3-D diagram of the cymbal transducer.

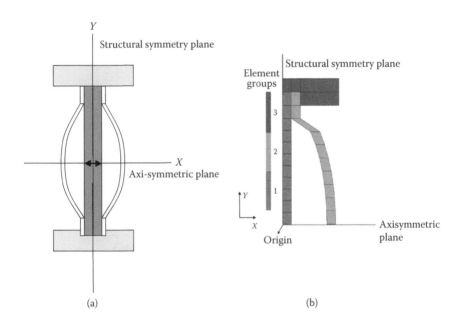

FIGURE 6.25
(a) The structural symmetry plane and the axi-symmetric plane of the transducer. (b) The axi-symmetric FE model of half the structure.

For the analysis of the transducer in water, acoustic FE elements and acoustic boundary elements are used to model the water body surrounding the transducer.

6.5.4.3.1.2 Material Properties and Dimensions The PZT SP-5A (Sparkler Ceramics Pvt. Ltd., Pune, India) is used for the piezoelectric discs and steel is used for the end-caps. The mount is made of rubber. Following are properties of the materials.

Compliance constant:

$$\begin{bmatrix} 16.4 & -5.74 & -7.22 & 0 & 0 & 0 \\ -5.74 & 16.4 & -7.22 & 0 & 0 & 0 \\ -7.22 & -7.22 & 18.8 & 0 & 0 & 0 \\ 0 & 0 & 0 & 47.5 & 0 & 0 \\ 0 & 0 & 0 & 0 & 47.5 & 0 \\ 0 & 0 & 0 & 0 & 0 & 44.3 \end{bmatrix} \times 10^{-12} \ m^2 / N$$

Piezoelectric d-coefficient:

$$\begin{bmatrix} 0 & 0 & 0 & 0 & 584 & 0 \\ 0 & 0 & 0 & 584 & 0 & 0 \\ -171 & -171 & 374 & & & \end{bmatrix} \times 10^{-12} \ C/N$$

Permittivity:

$$\begin{bmatrix} 15.31 & 0 & 0 \\ 0 & 15.31 & 0 \\ 0 & 0 & 15.045 \end{bmatrix} \times 10^{-9} \ F/m$$

Properties of steel and rubber:

Material	Density (kg/m³)	Elastic Constant (GPa)	Poisson's Ratio
Steel	7750	200	0.33
Rubber	1300	4	0.40

Dimensions:

PZT discs—Diameter, 25 mm

Thickness, 1 mm

Steel caps—Thickness, 1 mm

Cavity depth, 4 mm

Boundary conditions: The common face of the two discs is grounded.

6.5.4.3.2 Solving

6.5.4.3.2.1 Modal Analysis   Modal analysis of the transducer is carried out to obtain the natural frequencies.

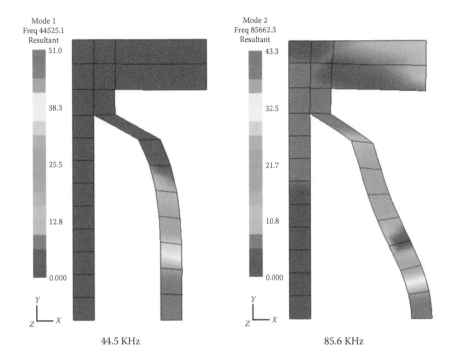

FIGURE 6.26
First two natural frequency mode shapes at 44.5 kHz and 85.6 kHz.

6.5.4.3.2.2 Harmonic Analysis Harmonic analysis is carried out by applying sinusoidal voltage of 1 V to the top and bottom faces of the PZT disc pair.

6.5.4.3.3 Postprocessing

6.5.4.3.3.1 Modal analysis

The first natural frequency: 44.5 kHz
The second natural frequency: 85.6 kHz
The two mode shapes are shown in Figure 6.26.

6.5.4.3.3.2 Harmonic Analysis in Air The graph of admittance as a function of frequency for the transducer in air in the frequency range of 40–90 kHz is shown in Figure 6.27. Two admittance peaks are observed, one at 44.2 kHz and the other at 83.4 kHz. The admittance values of the two frequencies are 0.031 Siemens and 0.230 Siemens, respectively.

6.5.4.3.3.3 Harmonic Analysis in Water The water body surrounding the transducer is modelled using acoustic FE and BE elements. The thickness of the acoustic FE on top of the structure is taken close to $(\lambda_{water}/4)$, where λ_{water} is the wavelength of sound in water at the highest frequency used in the analysis. The FE model of the transducer in water with FE and BE elements is shown in Figure 6.28.

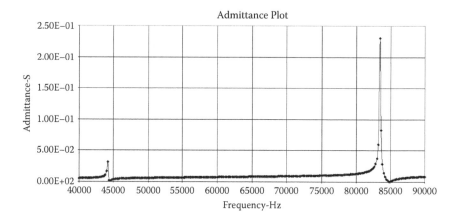

FIGURE 6.27
Admittance as a function of frequency for the cymbal transducer in air.

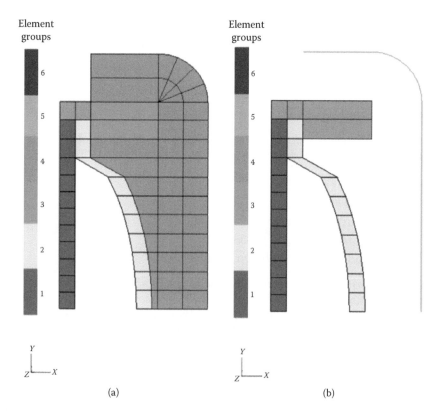

FIGURE 6.28
FE model of cymbal transducer in water. (a) Transducer with acoustic FE and BE elements. (b) Transducer with only BE elements shown.

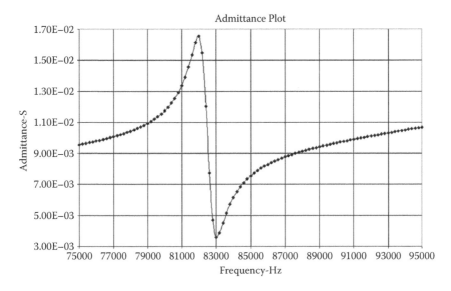

FIGURE 6.29
Admittance as a function of frequency for the cymbal transducer in water.

Admittance of the transducer in water as a function of frequency when 1 V is applied to the PZT discs is shown in Figure 6.29. No admittance peak is observed near the lower frequency of 44.2 kHz that was observed in air. A peak is observed at 82.0 kHz. The admittance value at 82.0 kHz is 0.0166 Siemens.

To find the distribution of pressure in water around the transducer acoustic pressure display elements are added as shown in Figure 6.30.

From the pressure display elements, the pressure values at various points in water near the transducer can be obtained. The pressure at a distance of 0.0156 m from the transducer cap top is found to be 5.2 kPa when a voltage of 1 V and frequency of 82.2 kHz is applied to the transducer.

Element
groups

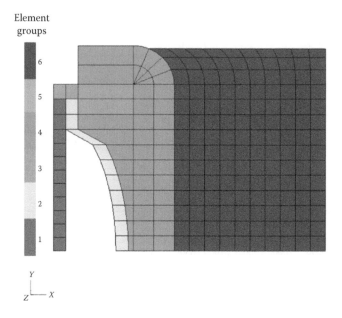

FIGURE 6.30
FE model of cymbal transducer in water with pressure display elements added.

References

1. T. R. Chandrupatla and A. D. Belegundu, 2002, *Introduction to Finite Elements in Engineering*, 3rd ed., Pearson, New Delhi.
2. R. Lerch, 1990, Simulation of piezoelectric devices by two- and three-dimensional finite elements, *IEEE Transactions on Ultrasonics, Ferroelectrics and Frequency Control*, Vol. 37, No. 3, 233–247.
3. F. Lowrie et al., FEM of electroceramics, NPL Report CMMT(A) 150.
4. http://www.efunda.com/materials/piezo/material_data/.
5. D. H. Johnson and D. Pal, 2000, Simulation of an ultrasonic piezoelectric transducer, http:/www.ohiocae.com/piezo-paper.htm. Proc. of the 9th Int. ANSYS Conf., Aug. 29, 2000, Pittsburg, PA.
6. X. Bao, Y. Bar-Cohen, Z. Chang, B. P. Dolgin, S. Sherrit, D. S. Pal, S. Du, and T. Peterson, 2003, Modeling and computer simulation of ultrasonic/sonic driller/corer (USDC), *IEEE, Transactions on Ultrasonics, Ferroelectrics and Frequency Control (UFFC)*, Vol. 50, No. 9, 1147–1160.
7. PAFEC-FE User Manual 8.8, PAFEC Ltd., UK.
8. P. W. Loveday and C. A. Rogers, 1998, Modification of piezoelectric vibratory gyroscope resonator parameters by feedback control, *IEEE Transactions on Ultrasonics, Ferroelectrics and Frequency Control*, Vol. 5, No. 5, 1211–1215.
9. PAFEC Acoustics User Manual Level 8.8.
10. L. E. Kinsler et al., 2000, *Fundamental of Acoustics*, 4th ed., John Wiley & Sons, Singapore.

Bibliography

1. *Advanced Piezoelectric Materials: Science and Technology*, 2010, Editor: K. Uchino, Woodhead Publishing.
2. *Piezoelectric Ceramic Materials: Processing, Properties, Characterization and Applications* (Materials Science and Technologies), 2010, Editor: W. G. Nelson, Nova Science Publishers.
3. *Piezoelectric Materials: Advances in Science, Technology and Applications*, 2008, Editors: C. Galassi, M. Dinescu, K. Uchino, Springer.
4. *Piezoelectricity: Evolution and Future of a Technology*, 2008, Walter Heywang, Karl Lubitz, and Wolfram Wersing, Springer.
5. *An Introduction to the Theory of Piezoelectricity*, Series: Advances in Mechanics and Mathematics, 2005, Vol. 9, Yang Jiashi, Springer.
6. *Electro Ceramics*, 2nd edition, 2003, A. J. Moulson and J. M. Herbert, John Wiley & Sons.
7. *Piezoelectric and Acoustic Materials for Transducer Applications*, 2008, Editors: Safari, Ahmad, Akdogan, E. Koray, Springer, NY, USA.

Index